MATHEMATICA  DISCOVERY

Andrew M. Bruckner University of California, Santa Barbara

Brian S. Thomson Simon Fraser University

Judith B. Bruckner ClassicalRealAnalysis.com

CLASSICALREALANALYSIS.COM

This text is intended for a course introducing the idea of mathematical discovery, especially to students who may not be particularly enthused about mathematics as yet.

Cover: Cover design by Karen Bruckner, based on an original graphic design by David Sprecher.

Citation: *Mathematical Discovery [Color Edition]*, Andrew M. Bruckner, Brian S. Thomson, and Judith B. Bruckner, ClassicalRealAnalysis.com (2011), xv 253 pp.
ISBN-13: 978-1463730574
ISBN-10: 1463730578

Date PDF file compiled: September 24, 2011

ISBN-13: 978-1463730574
ISBN-10: 1463730578

CLASSICALREALANALYSIS.COM

Contents

List of Figures

Preface

heu·ris·tic [adjective]

1. serving to indicate or point out; stimulating interest as a means of furthering investigation.

2. encouraging a person to learn, discover, understand, or solve problems on his or her own, as by experimenting, evaluating possible answers or solutions, or by trial and error: a heuristic teaching method.
   [Source: Dictionary.com]

Introduction

This book is an outgrowth of classes given at the University of California, Santa Barbara, mainly for students who had little mathematical background. Many of the students indicated they never understood what mathematics was all about (beyond what they learned in algebra and geometry). Was there any more mathematics to be discovered or created? How could one actually discover or create new mathematics?

In order to give these students some sort of answers to such questions, we designed a course in which the students could actually participate in the discovery of mathematics. The class was not presented in the usual lecture fashion. And it did not deal with topics that the students had seen before. Ordinary algebra, geometry, and arithmetic played minor roles in most of the problems we addressed. Whatever algebra and geometry that did appear was relatively easy and straightforward.

Our objective was to give the students an appreciation of mathematics, rather than to provide tools they would need in some field that required mathematics. In that sense, the course was like a course in music appreciation or art appreciation. Such courses don't attempt to train students to become pianists, composers, or artists. Instead, they attempt to give the students a sense of the subject.

Why do so many intelligent people have so little sense of the field of mathematics? A partial explanation involves the difficulty in communicating mathematics to the general public. Without special training in astronomy, medicine, or other scientific areas, a person can still get a sense of what goes on in those areas just by reading newspapers. But this is much more difficult in mathematics. This may be so because much of modern mathematics involves very technical language that is difficult to express in ordinary English. Even professional mathematicians often have difficulty communicating their work to other professional mathematicians who work in different areas.

This isn't surprising when one realizes how many areas and sub areas there are in mathematics. *Mathematical Reviews* (MR) is a journal that provides short reviews of mathematical papers that appear in over 2000 journals from around the world. The subject classification used by MR has over 50 subject areas, each of which has several subareas. Each of these subareas has many sub-sub areas. A research mathematician might be an expert in several of the sub-sub areas, be conversant in several areas, and know very little about the other areas.

Objectives

Our objective is to impart some of the flavor of mathematics. We do this in several ways. First, by actively participating in the discovery process, a reader will get a sense of how mathematicians discover new mathematics.

A problem arises. Discovery often begins with some experimentation to help give a sense of what is involved in the problem. After a while one might have enough understanding of the problem to be able to make a plausible conjecture, which one then tries to prove. The attempt to prove the conjecture can have several different outcomes. Sometimes the proof works. Other times it doesn't work, but in trying to prove it one learns much more about the problem and identifies some stumbling blocks.

Sometimes these stumbling blocks seem insurmountable and one tries to prove they actually *are* insurmountable—the conjecture is false. That may create its own stumbling blocks. All the time one learns more and more about the problem. Finally one either proves the conjecture or disproves it. (Or simply gives up!).

We shall see all of this unfolding in the several chapters in the book. Our discovery process will be similar to that of a research mathematician's, though our problems will be much less technical.

The first part of each chapter deals with a problem we wish to consider. We then go into the discovery mode and eventually obtain some answers. After this we turn to other aspects of mathematics related to the material of the chapter. What is the history of the problem? Who solved it? What are some related problems? How can other areas of mathematics be brought to bear on the problem? Do computers have any role in solving the problems raised? What about conjectures that seemed to be true, but were eventually proven false? Or remain unsolved?

We have tried to find some balance between discovery and instruction. This is not always possible: it is impossible to resist the many occasions when some idea leads naturally to another wonderful idea. The reader will not discover the connection, even with prodding, so we drop our heuristic approach and explain the new ideas. This is probably in the nature of things. When we look back on everything we have learned, certainly it is all a combination of stuff we figured out for ourselves and other stuff that we learned from others. It is the combination of the two that makes learning rewarding and productive. It is likely the stress on just the instruction part that explains the many people in this world who claim to dislike or fear mathematics.

Prerequisites

The main prerequisite for getting much from this book is curiosity and a willingness to attempt the problems we present. These problems usually set things up for the next stage in the discovery process. This is different from most text books, where the problems at the end of a section are intended to firm the readers' knowledge of the material just presented.

Almost all problems have answers supplied at the end of the chapter. The word ANSWER following a problem indicates that an answer is supplied. For readers using a PDF file on a computer or laptop screen, that word is hyperlinked to the answer. Readers working on a paperback version will have to scan the end of the chapter to find the appropriate answer.

When the book is read in a self-study manner, rather than in a classroom setting with an instructor to set the pace, there may be a temptation to move ahead quickly, to get to the end of the process to learn the result. (Did the butler commit the crime?). We urge that one resist the temptation. The students who got the most out of the class were the ones who participated actively in the discovery process. This included working the problems as they arose. They said that understanding this process was of more value to them than learning the answer.

In order to understand the material in most of the chapters, one needs a bit of algebra (just enough to be able to manipulate some simple algebraic expressions, though such manipulations play only a very minor role), a bit of geometry, and a little arithmetic.

One topic that is not usually covered in a first course in algebra is *mathematical induction*. This tool appears in several places. Readers not familiar with mathematical induction can reasonably work through a chapter that has an induction argument until that argument is needed. At that point, one can consult the Appendix where induction is discussed and induction proofs are given that are relevant to various problems we discuss. Induction does not take part in the discovery process—it is used only to verify that certain statements are true.

Rigor versus intuition

Professional mathematicians must be rigorous in their work. This involves giving careful definitions, even of apparently familiar objects. This often involves a great deal of "technical machinery." A mathematician needs to know such things as *exactly* what a "curve" is, what it means to "go around a curve so that the inside is to the left," how to mathematically describe the number of "holes" in a pretzel and the meaning of area.

It should be understood, however, that this is not the situation when a mathematician first starts thinking of a problem and working out a solution. Things are rather vague and intuitive in the early stages. The polish and rigor appear in full force only in the final drafts.

Since this book is not intended for mathematicians, who would require formal definitions and proofs, we can relax these requirements considerably. Everything we say in an informal way *can* be said in a mathematically rigorous way, but that is not our purpose. Our purpose is to provide some of the flavor of mathematics and introduce the reader to topics that some students were surprised to find involved mathematics. Thus we can take for granted that readers intuitively understand concepts such as curves, inside, left, holes, and area. We will occasionally describe a concept with which the reader may not be familiar, but our overall style is primarily a leisurely, informal one.

Acknowledgements

The authors are grateful to many who have contributed in one way or another to the creation of this book. In particular, the many students who participated in the classes contributed greatly. They pointed out difficulties, found interesting solutions to problems that were different from our solutions, provided reasonable approaches to problems that sometimes worked and sometimes didn't, but were often creative whether or not they worked.

We also want to acknowledge Professor Steve Agronsky who used preliminary versions of this book in his classes at Cal. Poly. University in San Luis Obispo, California. He made a number of useful suggestions based on his teachings .

Most of the figures were prepared by us using *Mathematica*.™ A number of the figures were found on the internet and, naturally enough, it has proved to be difficult to give proper attribution. Any person who is the original author of such figures is invited to write to us with instructions as to whom to give proper credit. We are thankful for those who have released such figures to the public domain and will be equally thankful to those who wish credit.

We are always grateful for comments and will attempt to incorporate them into future printings (or future editions) with explicit acknowledgement of the sources. Please write to the authors at thomson@sfu.ca.

To the Instructor

One might notice that, on occasion, one or more problems follow after only a short discussion. This occurs when we believe this short discussion already presents an opportunity for the reader to get a sense of how we might continue. When we taught the class, we often found it convenient to make a small amount of progress on each of two chapters in one class session. How this worked in practice varied with what happened in class discussion. Sometimes the material we list as problems actually became part of the class discussion, rather than as problems to be discussed at the next class session. It worked best to be flexible and see where the discussion took us in determining whether we should solve some of the problems in lecture form, or leave them as problems to be discussed in the next class meeting.

In a typical one-quarter term we would have covered four chapters in a leisurely fashion, at least through the discovery of the solution to the main problems of the chapter. We also were able to cover some of the material at the end of the chapters. Available time, class interests, and level of difficulty relative to the students' backgrounds determined what we covered.

We provide answers to most of the problems, in particular to those that point the way to further progress. We leave a few unanswered. Some of these we used as quizzes or homework to be collected.

"Perhaps I can best describe my experience of doing mathematics in terms of a journey through a dark unexplored mansion. You enter the first room of the mansion and it's completely dark. You stumble around bumping into the furniture, but gradually you learn where each piece of furniture is. Finally, after six months or so, you find the light switch, you turn it on, and suddenly it's all illuminated. You can see exactly where you were. Then you move into the next room and spend another six months in the dark. So each of these breakthroughs, while sometimes they're momentary, sometimes over a period of a day or two, they are the culmination of—and couldn't exist without—the many months of stumbling around in the dark that preceed them."

"I used to come up to my study, and start trying to find patterns. I tried doing calculations which explain some little piece of mathematics. I tried to fit it in with some previous broad conceptual understanding of some part of mathematics that would clarify the particular problem I was thinking about. Sometimes that would involve going and looking it up in a book to see how it's done there. Sometimes it was a question of modifying things a bit, doing a little extra calculation. And sometimes I realized that nothing that had ever been done before was any use at all. Then I just had to find something completely new; it's a mystery where that comes from. I carried this problem around in my head basically the whole time. I would wake up with it first thing in the morning, I would be thinking about it all day, and I would be thinking about it when I went to sleep. Without distraction, I would have the same thing going round and round in my mind. The only way I could relax was when I was with my children. Young children simply aren't interested in Fermat. They just want to hear a story and they're not going to let you do anything else."

— Andrew Wiles
In an interview for PBS TV program Nova on the occasion of his solving Fermat's Last Theorem.

Chapter 1

Tilings

It is easy to imagine a rectangle tiled with squares. The familiar checkerboard in Figure 1.1 is a tiling of a square by sixty-four smaller squares.

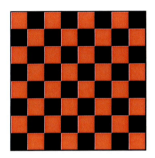

Figure 1.1: Checkerboard.

A little more artistically, the tiling in Figure 1.2 shows a rectangle that has been tiled into a number of smaller squares arranged in an attractive design.

Figure 1.2: Greek mosaic made with square tiles.

In both these cases all the squares are of equal size. This is familiar in the pattern we see for checkerboards or for many ceramic tilings of kitchen floors. But what if the squares are not all of the same size?

Figure 1.3 has tiles of unequal size but many of them are of the same size. What if we insist that *no two of the squares can be of the same size*. A few moments of thought shows that this problem is much, much harder.

How does one begin to discover such constructions? Perhaps after trying to find one you will give up in frustration and suspect that no such tiling can exist.

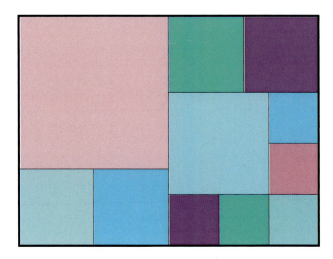

Figure 1.3: Tiling a rectangle with squares

We don't recognize this as a problem that we can attack by any of the standard methods of arithmetic, algebra or geometry. This is a situation that often arises in creative mathematics. We are faced with a problem but are at a loss about what tools to bring to bear on the problem. What to do? Faced with this type of problem, the creative mathematician would probably begin by trying to get a *feel* for the problem by experimenting with a few examples.

1.1 Squaring the rectangle

The problem of tiling a rectangle with unequal sized squares has been described by some as the problem of *squaring the rectangle*. We do not know in advance on starting to look at such a problem whether there is a solution, and if there is a solution how we should go about finding one.

Perhaps we should begin by seeing whether we can put together a few squares (no two of the same size) in such a way that they combine to form a rectangle. (At this stage, it's almost like working a jig-saw puzzle.)

Let's start with a small number of squares. A moment's reflection reveals that it is impossible to achieve our desired result with only two or three squares. With four squares, there are quite a few ways in which the squares can be combined. Figure 1.4 shows two possibilities that you might have tried.

Problem 1 *Experiment with four, five, and six squares. That is, try to combine the squares in such a way that the resulting figure is a rectangle. Remember that* no two squares can be the same size.   Answer □

Figure 1.4: Tiling a rectangle with four squares?

1.1.1 Continue experimenting

Did you find a tiling of a rectangle by four, five, or six squares, all of different sizes? If so, check again. Are two of the squares the same size? You do not need a ruler to check this. Simply put in the numbers which you think represent the lengths of the sides of the squares and see if everything adds up right. For example, we might think that the configurations in Figure 1.4 are possible after all. Maybe our drawing program does not quite get the job done, but the configuration there is possible with the right choice of dimensions.

The chances are that you did not arrive at a solution to the problem. It must also have become clear that as the number of squares we use in our experimenting increases, the number of essentially different configurations we can put together increases rapidly. Even with six squares, the number of configurations we can try is very large—and it gets much worse if we tried to use seven or eight tiles.

How should we proceed? Our experimenting has not brought us a solution to the problem. But that does not mean it was a waste of time. We may have learned something.

1.1.2 Focus on the smallest square

For example, we may have noticed that many of our attempts led to a certain difficulty. Perhaps we can find a way to overcome this difficulty. Or, perhaps it is impossible to overcome, thereby making the problem one with no solution. What is this difficulty? Consider again, for a moment, the configurations that you tried out while working on Problem 1. For each of these look to see where you placed the smallest square.

In each case there appeared a small space neighboring the smallest tile. Perhaps you noticed a similar state of affairs in some of your attempts with four,

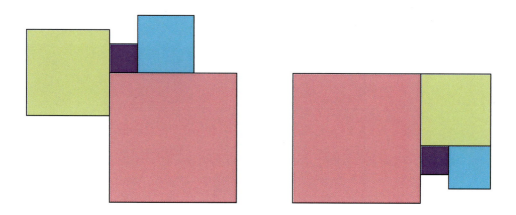

Figure 1.5: Where is the smallest square?

five or six tiles. If we were able to complete these attempts by adding more tiles, these small spaces could accommodate only tiles which are small enough to fit into the space. And this would create even smaller spaces, to be filled with even smaller tiles. We can certainly continue to add smaller and smaller tiles, but at some point the process must stop if we are to arrive at a solution to our problem. At this point it may look hopeless. Perhaps we can use what we have learned to prove that there is no solution to the problem that uses only four, five or six squares.

1.1.3   Where is the smallest square

Let us focus on the difficulty we encountered. If there *is* a solution, there must be a smallest square S. And that smallest square S must fit into the picture somewhere. Where? Maybe we can show that there is no place for it to fit.

This is what our experimenting showed — whenever the smallest square was in one of our trials, there was a space neighboring it which could accommodate only still smaller squares. (This might not have been true of all our trials, but it probably was true of most of those trials that offered any hope of success.) Where could the smallest square fit? Could it be in a corner as in Figure 1.6?

Figure 1.6: Where is the smallest square? (In a corner?)

Is the smallest square in a corner? A moment's reflection shows it can't be. Since S is the smallest square, its neighbors must be larger as in Figure 1.7.

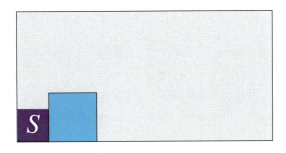

Figure 1.7: The smallest square has a larger neighbor.

But that creates exactly the kind of space we've been talking about. Only squares smaller than S could fit into that space.

Is the smallest square on a side? Similarly, we see that S cannot be on one of the sides of the rectangle as Figure 1.8 illustrates.

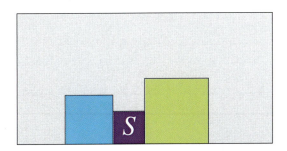

Figure 1.8: The smallest square has two larger neighbors.

It's two neighbors on that side must be larger than S; once again a small space is created. So, if there is a solution to the problem at all, the smallest square must lie somewhere inside the rectangle, i.e., its sides cannot touch the border of the rectangle.

Problem 2 *Do you think it is possible to find a tiling using exactly four squares of unequal size?* □

Problem 3 *Do you think it is possible to find a tiling using exactly five or six squares of unequal size?* □

1.1.4 What are the neighbors of the smallest square?

Did you find a tiling with five or six squares? If so, you'd better check that it really works. Did you find a proof that there is no solution? If so, you'd better make sure you really have a proof.

Let's analyze a bit more. Suppose there is a solution and S is the smallest square. We know S must be inside the rectangle. What possibilities are there for the relationship between S and its neighbors?

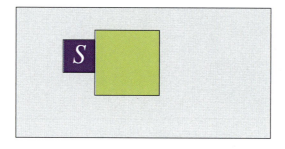

Figure 1.9: Possible Neighbor of the smallest square? (No.)

A possible case?   A neighbor of S might extend beyond S *on both sides* as Figure 1.9 illustrates. This, we see is not possible because two other neighbors (the ones below and above S in the diagram) would then create a small space.

Another possible case?   The smallest square S may have a side bordering on two neighbors as Figure 1.10 illustrates. This is impossible for the same reason.

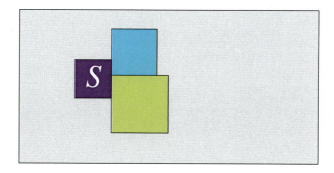

Figure 1.10: Two possible neighbors of smallest square? (No.)

The only possible case!   Each neighbor of the smallest square S has a side which fully contains one side of S, but extends on one side of S only. Figure 1.11 illustrates this. Is this possible? At least no small space has been created. This is the only case we cannot rule out immediately.

What does a solution look like?   We now know that if there is a solution, the only possible placement of the smallest square S is that S be somewhere inside the rectangle and be surrounded by its neighbors in a windmill fashion.

Figure 1.11: Four possible neighbors of smallest square? (Maybe.)

We have not determined that a solution exists. But we have learned something about what a solution must look like (if there is a solution at all).

This leaves us with two options: we could continue to try to show there is no solution. How might we try? Perhaps we can still show that there is no place to put S. Or maybe the second smallest square creates a problem. Our alternative is to switch gears again and try to show there is a solution. If we take this positive option, we are far better off than we were at the beginning. We need try only such constructions which have the smallest square surrounded by its neighbors in a windmill fashion. Let's try that for awhile and see what it leads to.

Problem 4 *Experiment with four, five, and six squares trying to combine the squares in such a way that the resulting figure is a rectangle. (Same as Problem 1, but use newly learned information.)*

Answer □

1.1.5   Is there a five square tiling?

It is clear that we need not try to find a solution with four squares. One thing we've already learned is that a solution (if one exists) requires at least five squares, namely S and its four neighbors. Let's try a solution with five squares. Such a solution must involve S surrounded by its neighbors in a windmill fashion. Figure 1.12 illustrates an attempt at this. In the figure A, B, C and D are squares surrounding a central square S.

Careful measurements of the sides of the squares in this configuration will reveal that they are not exactly squares. (And we want them exactly squares.) But that may mean no more than that we weren't careful with our drawing. And, after all, no one can draw a perfect square! One would hardly discard the idea of a circle just because no one can draw a perfect circle.

If we think the diagram above represents a solution, we should try to find

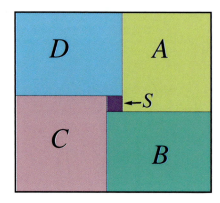

Figure 1.12: We try for a five square tiling.

numbers representing the sides of the squares so that all the requirements of our problem are satisfied.

An algebraic method   To check that a proposed solution is correct or to prove that a proposed solution is impossible, we can use some simple algebra. Suppose the diagram represented a solution. Denote the length of the side of S by s and the length of the side of A by the letter a. The labeling is shown in Figure 1.13.

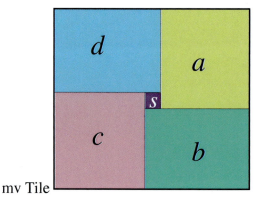

mv Tile

Figure 1.13: a, b, c, d, and and s are the lengths of the sides of the "squares."

Then, B has side length $s + a$ (why?) so C has side length

$$s + (s + a) = 2s + a$$

and D has side length

$$s + (2s + a) = 3s + a.$$

But, looking at A, S, and D, we see that $a = d + s$. Thus $a = 4s + a$, that is, $s = 0$. This shows that our configuration is impossible. The square S reduces to a point, and the other four squares are all of the same size.

The only other possible five-square configuration using our windmill idea would look similar to this and would check out negatively too. To this point, then, we have proved that it is impossible to solve our problem with five or fewer squares.

1.1.6   Is there a six, seven, or nine square tiling?

In the problems below determine whether the suggested configurations can work. Don't go by the accuracy of the drawing. Just because some of the tiles don't look like squares doesn't mean that one can't distort the picture some, keeping each tile in its same relationship to its neighbors, and making all the tiles squares. In some cases you may need to use the algebraic technique of this section.

Problem 5 *Does this configuration in Figure 1.14 of six "squares" work?*

Figure 1.14: A tiling with six squares?

Answer □

Problem 6 *Does the configuration of seven "squares" in Figure 1.15 work?*

Answer □

Problem 7 *Does the configuration of nine "squares" in Figure 1.15 work?*

Answer □

Problem 8 *Experiment some more. Construct diagrams like those in Problem 5, Problem 6 and Problem 7.*

Answer □

Figure 1.15: A tiling with seven squares? With nine squares?

1.2 A solution?

While working on Problem 8 you may have succeeded in arriving at a diagram such as the one that appears in Figure 1.16. We don't have to sketch it accurately; the figure suggests another possible configuration that might look like this. As usual, for our method, the smallest square is labeled as s and its neighbor as a. The rest of the side lengths would then be determined as the figure shows.

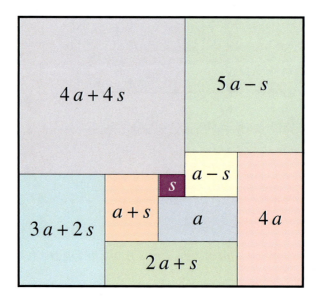

Figure 1.16: Will this nine square tiling work?

Can this configuration be made into a solution? That is, can values of s and a be found so that all the rectangles are squares? Since the right and left sides of the rectangle must have the same length, we calculate

$$7a + 6s = 9a - s$$

or

$$7s = 2a.$$

If, for example. we take $a = 7$ and $s = 2$ we would have $7s = 2a$ and we would arrive at the following diagram in Figure 1.17, the tiny square having side 2.

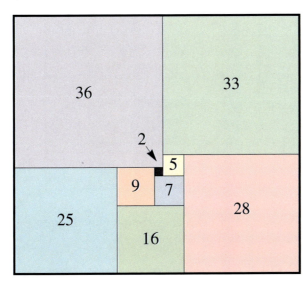

Elements: 2, 5, 7, 9, 16, 25, 28, 33, 36

Figure 1.17: A tiling with nine squares!

Thus, we see there is a solution to the nine square problem after all. And, to be sure, the diagram that we and you used for this solution would not have had tiles that *looked like squares* (unlike the final neat graphics here) but the algebra verified that we can create a tiling meeting all our conditions.

Problem 9 *Here is the algebraic method of this section as described by William T. Tutte (1917–2002), one of the founders of this theory:*

> "The construction of perfect rectangles proved to be quite easy. The method used was as follows. First we sketch a rectangle cut up into rectangles, as in [Figure1.18]. We then think of the diagram a bad drawing of a squared rectangle, the small rectangles being really squares, and we work out by elementary algebra what the relative sizes of the squares must be on this assumption. Thus in [Figure1.18] we have denoted the sides of two adjacent small squares by x and y and then that the side of the square next on the left is $x + 2y$, and so on. Proceeding in this way we get the formulae ... for the sides of the 11 small squares. These formulae make the squares fit together exactly .... This gives the perfect rectangle ... the one first found by [Arthur] Stone." ——*W. T. Tutte [12].*

Carry out all the arithmetic needed to construct Figure 1.18, the initial sketch for Stone's tiling. Then do the necessary algebra to find the sides of the eleven squares. Answer □

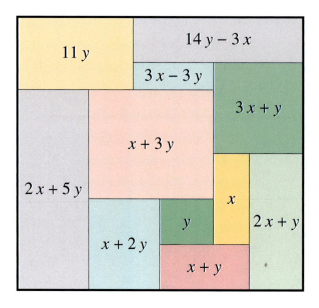

Figure 1.18: Initial sketch for Arthur Stone's eleven-square tiling.

1.2.1   Bouwkamp codes

Our solution of the rectangle in Figure 1.17 tiled with nine squares is something we might want to keep a record of and communicate to others. If we send someone a picture they can easily check that we have it all right and can see exactly what our solution is. Suppose we communicate only the size of the smaller squares:

$$2, 5, 7, 9, 16, 25, 28, 33, 36.$$

A little more helpful would be to indicate also the size of the large rectangle, in this case

$$61 \times 69.$$

In theory that should be enough for someone who likes fiendish puzzles, but these numbers alone don't tell the story in any adequate way. The picture does, but that is an inefficient way to communicate our ideas.

The Dutch mathematician Christoffel Jacob Bouwkamp (1915–2003) devised a simple code that is much used nowadays. Problem 10 asks you to devise your own code, but the answer (found at the end of the chapter) gives the Bouwkamp code and a brief description of how it works.

Problem 10 *There are 21 square tiles in Figure 1.19. How could you send a text message to a friend (no pictures allowed) that would allow him to reconstruct this tiling?*

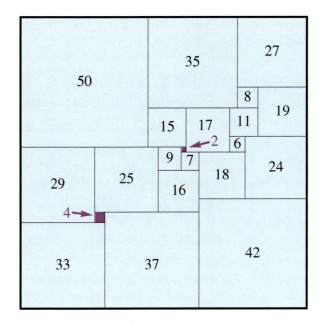

Figure 1.19: Can you reconstruct this figure from the numbers?

Answer □

Problem 11 *Give the Bouwkamp code for Figure 1.17.* Answer □

Problem 12 *Here are the Bouwkamp codes for the only ninth order squared rectangles. Construct the one that is not in the text already.*

```
Order 9, 33 by 32: (18,15)(7,8)(14,4)(10,1)(9)
Order 9, 69 by 61: (36,33)(5,28)(25,9,2)(7)(16)
```

Answer □

1.2.2 Summary

Let us reflect on where we have been so far in this chapter. We started with an interesting (but puzzling) geometric problem. It was unlike the usual high-school geometry problems in that none of the usual techniques of geometry could be brought to bear on the problem.

At first, the problem wasn't one for which we had any ideas at all for a solution. So we played around with it in the hopes of learning something. What we learned by experimenting enough was that there was a difficulty caused by the small space adjoining the smallest square in most of our attempts. Maybe that was the key to the problem. Perhaps there was no solution, and perhaps we could prove that by showing there's no place for the smallest square.

We succeeded in eliminating certain placements for the smallest square, seeing that such placements always created a *small space* that needed an even smaller square. But one such placement did not seem to lead to any problem.

We returned to the drawing boards, armed with our new information. Eventually we were able to use a bit of algebra together with what we learned to arrive at a solution.

So we've solved our problem. Now what? A creative mathematician might ask a lot of questions suggested by this problem. Which rectangles can one tile with squares? Are there any squares that can be tiled with unequal squares? What other tilings are possible or impossible?

For additional examples of tilings, see Stein [9]. In that reference one can find a leisurely development of a number of questions related to tiling. In particular, a surprising way in which tiling and electrical theory are related is developed there and leads to the theorem that if a rectangle can be tiled with squares in any manner whatsoever, then it can also be tiled by squares all of the same size.

We will continue with some related material for those readers who want to pursue these ideas further. For mathematicians no problem ever stops cleanly: there are always some more questions to address, more ideas that our investigation suggests.

1.3   Tiling by cubes

What about tilings with other types of figures? One can ask analogous questions in higher dimensions. Is it possible to fill a rectangular box with cubes no two of which are the same size (as suggested in Figure 1.20)? This is the three dimensional version of the problem we just solved. At first glance it appears to be much more difficult. But, perhaps some of the insights we picked up from the two-dimensional case can be of use to us in this three dimensional version.

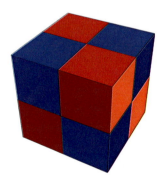

Figure 1.20: Tiling a box with cubes.

Problem 13 *Determine whether or not it is possible to fill a three-dimensional rectangular box with cubes, no two of which are the same size.*   Answer □

1.4 Tilings by equilateral triangles

Figure 1.21 shows a tiling of an equilateral triangle with other equilateral triangles, but notice that there are several duplications of same sized triangles in the figure.

Figure 1.21: Equilateral triangle tiling.

Similar ideas to those developed so far in the chapter are useful in showing that it is impossible to tile an equilateral triangle with other equilateral triangles no two of which are the same size. Problem 14 asks you to do this.

1.4.1 (Tutte, 1948) *If an equilateral triangle is tiled with other equilateral triangles then there must be two of the smaller triangles of the same size.*

This was first proved by W. T. Tutte in 1948 (see item [11] in our bibliography). An accessible account of this problem appears as the chapter

W. T. Tutte, *Dissections into equilateral triangles* (pp. 127–139)

in the book by David Klarner that is reference [16] in our bibliography. A 1981 article by Edwin Buchman in the American Math. Monthly (see [15]) shows, using Tutte's methods, that there is no convex figure at all that could be tiled by equilateral triangles unless at least two of those triangles are the same size.

For further discussion of these topics see the book of Sherman Stein [9] that appears in our bibliography.

Problem 14 *Show that it is not possible to tile an equilateral triangle with smaller equilateral triangles, no two of which are the same size.*   Answer □

1.5   Supplementary material

We conclude our chapter with some supplementary material that the reader may find of interest in connection with the problem of squaring the rectangle.

1.5.1   Squaring the square

We have succeeded in tiling some rectangles with unequal squares but none of our rectangles was a square itself. Is it possible to assemble some collection of unequal squares into a *square*?

The description of the problem as *squaring the square* originates with one of the four Cambridge University students Tutte, Brooks, Smith, and Stone who attacked the problem in 1935. It was intended humorously since it seems to allude to the famous problem of *squaring the circle* which means something totally different and was well-known to be impossible.

Tutte in his autobiographical memoir[1] describes Arthur H. Stone (1916-2000) as the one of the four who proposed the problem. He had found an old puzzle in a book of Victorian puzzles written by Henry Dudeney, an English puzzler and writer of recreational mathematics.

Figure 1.22: Tutte and Stone.

See Figure 1.23 for Dudeney's statement of his problem. The "solution" of the problem in the book is given by Dudeney in Figure 1.24 where the inlaid strip of gold is the black rectangle in the middle. The problem is called *Lady Isabel's Casket*. (In Victorian England a casket was not necessarily just for containing corpses, but could be "a small box or chest, often fine and beautiful, used to hold jewels, letters or other valuables" [as defined in the World Book Dictionary].)

---

[1]*Graph Theory As I Have Known It*, by W. T. Tutte (item [13] in our bibliography).

40.—*Lady Isabel's Casket.*

Sir Hugh's young kinswoman and ward, Lady Isabel de Fitzarnulph, was known far and wide as "Isabel the Fair." Amongst her treasures was a casket, the top of which was perfectly square in shape. It was inlaid with pieces of wood and a strip of gold, ten inches long by a quarter of an inch wide.

When young men sued for the hand of Lady Isabel, Sir Hugh promised his consent to the one who would tell him the dimensions

of the top of the box from these facts alone : that there was a rectangular strip of gold, ten inches by $\frac{1}{4}$-inch ; and the rest of the surface was exactly inlaid with pieces of wood, each piece being a perfect square, and no two pieces of the same size. Many young men failed, but one at length succeeded. The puzzle is not an easy one, but the dimensions of that strip of gold, combined with those other conditions, absolutely determines the size of the top of the casket.

Figure 1.23: Lady Isabel's Casket (from a 1902 English book of puzzles).

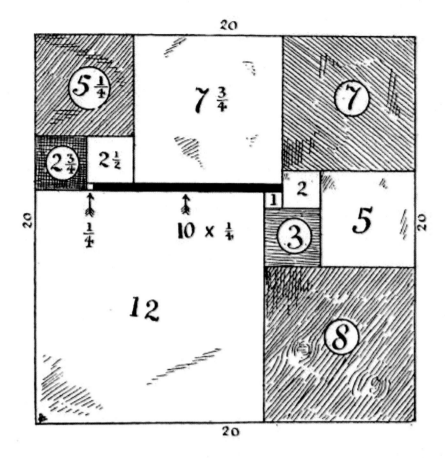

Figure 1.24: The "solution" to Lady Isabel's Casket.

Stone realized that the problem was tougher than Dudeney had thought, for, if this figure were indeed the *unique* solution of the problem, that could only mean that none of the squares in the figure could be divided into smaller unequal squares. They learned that the great Russian mathematician Nikolai Nikolaievich Lusin (1883–1950) had conjectured that no square could be squared. Thus the four of them decided that they could make their reputation by solving this Lusin Conjecture: *no square can be subdivided into a collection of squares no two of the same size.*

In fact they not only succeeded in squaring the square but in finding deep connections to the problem with graph theory and electrical networks.

The smallest squared-square   Did you notice that Figure 1.19 is a squared-square? Problem 10 asked for the Bouwkamp code for this tiling by twenty-one unequal squares. This is the lowest order example of squaring the square.

Observe that every square S, whether the length of its sides is an integer, a rational number or even an irrational number, can be tiled with squares of unequal size. Just shrink or stretch the square in Figure 1.19 to the size of S.

This gives a tiling of *S*.

A final word. The problem that began this chapter was to determine whether it is *ever* possible to tile a rectangle with squares of unequal sizes. We answered this question in the affirmative. The question remains *which* rectangles can be tiled in this manner. The answer to this question is given following the answer to Problem 19.

1.5.2 Additional problems

For those readers who did not get enough problems to work on here are some more. We also have added some more Bouwkamp codes problems as they appear to be popular entertainments (much like Suduko problems). Note that with these codes one can design jig-saw puzzles consisting of unequal squares which must be assembled to form a large rectangle. The Bouwkamp codes themselves then are quick descriptions of how to assemble the pieces to solve the puzzle.

Problem 15 *Here are the Bouwkamp codes for all of the tenth order squared rectangles. Sketch the tiling figures for as many of these as you find entertaining.*

```
Order 10, 105 by 104: (60,45)(19,26)(44,16)(12,7)(33)(28)
Order 10, 111 by 98:  (57,54)(3,7,44)(41,15,4)(11)(26)
Order 10, 115 by 94:  (60,55)(16,39)(34,15,11)(4,23)(19)
Order 10, 130 by 79:  (45,44,41)(3,38)(12,35)(34,11)(23)
Order 10, 57 by 55:   (30,27)(3,11,13)(25,8)(17,2)(15)
Order 10, 65 by 47:   (25,17,23)(11,6)(5,24)(22,3)(19)
```

□

Problem 16 *Here are the Bouwkamp codes for all of the eleventh order squared rectangles. If this still amuses you, sketch some more figures.*

```
Order 11, 112 by 81:   (43,29,40)(19,10)(9,1)(41)(38,5)(33)
Order 11, 177 by 176:  (99,78)(21,57)(77,43)(16,41)(34,9)(25)
Order 11, 185 by 151:  (95,90)(5,24,61)(56,25,19)(6,37)(31)
Order 11, 185 by 168:  (100,85)(43,42)(68,32)(1,41)(4,40)(36)
Order 11, 185 by 183:  (105,80)(33,47)(78,27)(19,14)(5,56)(51)
Order 11, 187 by 166:  (99,88)(10,78)(1,9)(67,25,8)(17)(42)
Order 11, 191 by 162:  (97,94)(26,68)(65,32)(9,17)(33,8)(25)
Order 11, 191 by 177:  (102,89)(40,49)(75,27)(48,19)(10,39)(29)
Order 11, 194 by 159:  (100,94)(29,65)(59,25,16)(9,7)(36)(34)
Order 11, 194 by 183:  (102,92)(31,23,38)(81,21)(8,15)(60)(53)
Order 11, 195 by 191:  (105,90)(15,31,44)(86,34)(18,13)(57)(52)
Order 11, 199 by 169:  (105,94)(19,75)(64,33,8)(27)(31,2)(29)
Order 11, 199 by 178:  (102,97)(16,81)(76,15,11)(4,23)(19)(42)
Order 11, 205 by 181:  (105,100)(6,13,81)(76,28,1)(7)(20)(48)
Order 11, 209 by 127:  (72,71,66)(5,61)(1,19,56)(55,18)(37)
Order 11, 209 by 144:  (85,57,67)(47,10)(77)(59,26)(7,40)(33)
Order 11, 209 by 159:  (89,49,71)(27,22)(5,88)(32)(70,19)(51)
Order 11, 209 by 168:  (92,64,53)(11,42)(44,31)(76,16)(73)(60)
```

```
Order 11, 209 by 177: (96,56,57)(55,1)(58)(81,15)(66,4)(62)
Order 11, 97 by 96: (56,41)(17,24)(40,14,2)(12,7)(31)(26)
Order 11, 98 by 86: (51,47)(8,39)(35,11,5)(1,7)(6)(24)
Order 11, 98 by 95: (50,48)(7,19,22)(45,5)(12)(28,3)(25)
```

□

Problem 17 *If a rectangle is tiled by squares, all of different sizes, the second smallest square cannot touch the border of the rectangle. Prove this statement.-*

□

Problem 18 *Suppose we are given a rectangle of dimensions a × b. Can this rectangle be subdivided into equal sized squares?*   Answer □

Problem 19 *Suppose we are given a rectangle of dimensions a × b. Under what circumstances can you be sure that this rectangle* cannot *be subdivided into a finite number of (not necessarily equal) squares?*   Answer □

1.6   Answers to problems

Problem 1, page 2

Figure 1.25 shows some possibilities with four unequal squares that you might have tried. These two are unsuccessful.

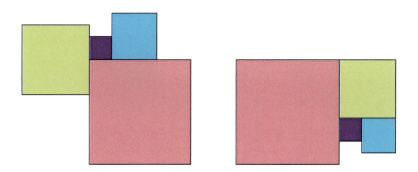

Figure 1.25: More experiments with four squares.

Problem 4, page 7

A four square configuration can't have a windmill and so we can pass over the possibility of a four square arrangement. Your first try at the five square configuration (using the windmill idea around the smallest square) might look like that in Figure 1.26. Does this, indeed, represent a possible solution *if we get the dimensions right*? Our drawing program won't produce accurate squares but the layout looks promising.

Before reading on in the text, try to see if you can find dimensions that would make this configuration work. Label the side lengths of the squares and see if there are numbers that work. If you can show that there cannot be such numbers then you will have succeeded in showing that this particular arrangement does not work.

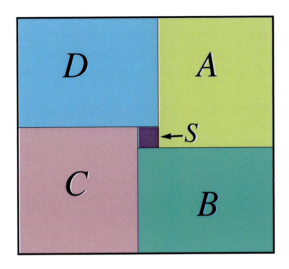

Figure 1.26: We try for a five square tiling.

Problem 5, page 9

In the figure of Problem 5, we see that two of the tiles have a common side. If they are to be squares, they must be the same size, violating a condition of our problem. Thus we do not need to do the algebra. A tiling that looks like this does not solve our primary problem: find a tiling with all squares of *different sizes*.

Problem 6, page 9

In the figure for Problem 6, we see a plausible configuration. None of the squares (if indeed they could be squares) is the same size as any of the others. We need to find exact numbers that would make this work.

If the diagram could be a solution, we can spot which of the squares could be the smallest. Denote the side of that square by s and denote the side of its right-hand neighbor by a. We then compute the (sizes of) the sides of the remaining squares arriving at the diagram in Figure 1.27.

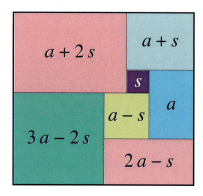

Figure 1.27: Lengths in terms of sides of 2 adjacent squares for Figure 1.15.

Since the top and bottom of a rectangle are of equal length,

$$3s + 2a = 5a - 3s$$

so that $2s = a$. Thus two of the rectangles would have to have sides equal to $3s$. Again this violates our primary objective: find a tiling with all squares of *different sizes*. Did you notice that other requirements are violated?

Problem 7, page 9

Again we see no immediate objection to this configuration: it might work. Let's do our algebraic computations. There are several ways to do this. Here's one in Figure 1.28. that gives us the sizes of some of the squares. We now compute that the darkest square in the figure has side

$$(a - 3s) - a - s = -4s.$$

This is again impossible, now because we have produced a negative number for the length of a side.

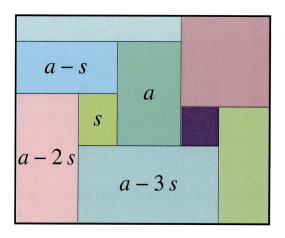

Figure 1.28: Some square lengths labeled for Figure 1.15.

Remark   Note that our arguments never involved a statement such as "this tile is much too thin to be a square." Even if we had been correct with such a statement, this would not rule out a similar configuration in which all the tiles were squares.

Such a configuration could possibly have been achieved by proper vertical and horizontal stretchings of the entire configuration. But our arguments in all three of the problems in this section showed that something was inherently wrong with the ways some of the tiles related to their neighbors. One couldn't stretch the configurations and render all the tiles squares of different sizes.

Problem 8, page 9

O.K. We are now ready to take another crack at finding a solution. We have a simple and easy to apply algebraic method for checking our proposed solution. We know in advance where to place the smallest square.

If our attempts fail, perhaps we can discover some unresolvable difficulty inherent in the problem. If we can prove that there is such an inherent unresolvable difficulty, then we will have proved the problem has no solution. Many problems posed in mathematics have no solution and we might be equally proud of showing that the problem is impossible as finding an answer.

But first, experiment some more.

Keep in mind that there must be more than five squares, the smallest must be surrounded by its neighbors in a windmill fashion, and the requirements of the problem must be met. Apply our algebraic method for obtaining the sizes of the sides of the tiles (if they are to represent a solution) and see what that leads to. Instead of proceeding almost blindly, try to modify diagrams you have already studied, such as those in this section. See what went wrong with these attempts and try to overcome the difficulty (or try to find some irreconcilable difficulty).

Problem 9, page 11

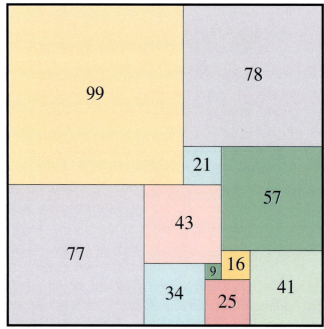

Elements: 9, 16, 21, 25, 34, 41, 43, 57, 77, 78, 99

Figure 1.29: Realization of Arthur Stone's eleven-square tiling.

The dimensions of Stone's tiling are shown in Figure 1.29. Just do the elementary algebra using the same method that we used and you should be able to discover all of the dimensions. You may wish to compare the length of the side of the rectangle in the upper right hand corner with the lengths of the sides of the two rectangles below it to obtain $16y = 9x$.

Problem 10, page 12

A reasonable start at communicating the configuration in Figure 1.19 is to start at the upper left corner and report the adjacent squares at the top from left to right:

$$50, 35, 27.$$

Then what to report next? You might decide to spiral around the outside of the square in a clockwise direction. But that would likely end up in trouble. The Bouwkamp method is just to keep reporting left to right all the new squares you see at each level. There are ten levels in the picture (count them) and so you need a report at each of these levels. The level is defined by the top of the squares, starting with the very top level which we decided to report by the numbers [50, 35, 27].

In the Bouwkamp code, brackets are used to group adjacent squares with flush tops, and then the groups are sequentially placed in the highest (and leftmost) possible slots. For this example of the 21-square illustrated in the problem the code is

$$[50,35,27], \ [8,19], \ [15,17,11], \ [6,24], \ [29,25,9,2], \ [7,18], \ [16], \ [42], \ [4,37], \ [33].$$

Problem 11, page 13

$$[36,33], \ [5,28], \ [25,9,2], \ [7], \ [16].$$

Problem 12, page 13

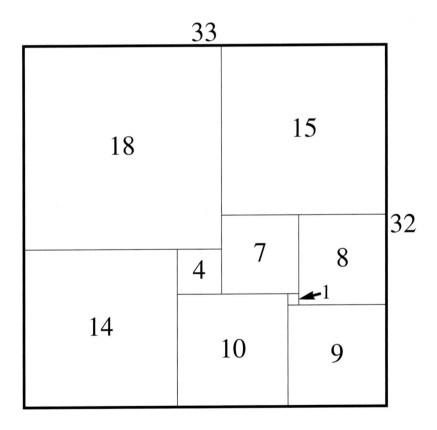

Figure 1.30: A 33 by 32 rectangle tiled with nine squares.

In Figure 1.30 is a picture that corresponds to the Bouwkamp code

```
Order 9, 33 by 32: (18,15)(7,8)(14,4)(10,1)(9).
```

Problem 13, page 14

It is a bit more difficult to experiment with the three-dimensional setting than it was with the two-dimensional setting. The remark before the problem suggests that "some of the insights we picked up from the two dimensional case can be of use to us in this three dimensional version." The key in two dimensions was the use of the smallest square argument. Try this:

Use the smallest cube argument!

Don't read the rest of the answer without trying again. You may wish to glance at Figure 1.31.

Our proof is an indirect one. We assume that there is such a construction and find that there is a contradiction.

Suppose a rectangular box were filled with cubes no two of which were of the same size. Consider only those cubes which lie on the bottom of the box. The bottom faces of these cubes tile the floor of the box by squares, no two of the same size. The smallest of these tiles must be surrounded by four other tiles in a windmill fashion. Let K_1 be the smallest of the cubes lying on the floor of the box. From what we just said, we see that K_1 is surrounded by four larger cubes which form a *tower* around K_1 as suggested in Figure 1.31.

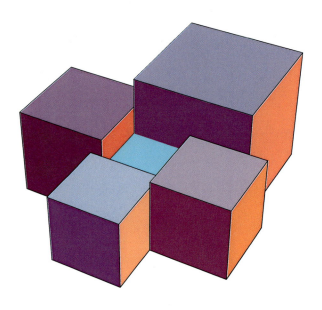

Figure 1.31: A tower of cubes around K_1.

Now consider those cubes whose bottoms lie on the top face of K_1. Their bottom faces tile the top face of K_1. As before, we conclude that the smallest of these K_2 is surrounded by four larger cubes which form a tower around it.

Continuing in this manner we see there can be no end to this process. No matter how many of these cubes K_1, K_2, K_3, ... we have obtained, there must still be smaller ones lying on top of the smallest obtained to that point.

Thus, we have proved this:

1.6.1 (No cubing the box) *It is impossible to fill a rectangular box with cubes, all of different sizes.*

Our techniques in the tiling problem of studying the location of the smallest square was useful to us in two ways: firstly, it gave us information about the structure of tilings of rectangles by squares of different sizes—the smallest square must be surrounded in a certain way by its neighbors; secondly. It suggested an approach to solving the analogous problem in three dimensional space.

Problem 14, page 15

The smallest cube argument that succeeded for Problem 13 suggests that a *smallest triangle argument* can be developed for this problem, and indeed very similar ideas will work here.

Our proof is again an indirect one. We assume that there is such a construction and find that there is a contradiction.

Assume that we have a tiling by smaller equilateral triangles, all of different sizes. Start by looking for the smallest triangle S that touches the bottom of the triangle. Argue that it must look like Figure 1.32.

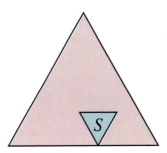

Figure 1.32: S is the smallest triangle at the bottom of the tiling.

Then look for the smallest triangle T that touches the top of the triangle S. Argue that it must look like Figure 1.33. This argument keeps going indefinitely and so we shall soon run out of triangles, just as in our solution to Problem 13 we ran out of cubes.

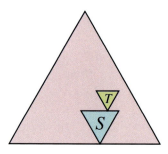

Figure 1.33: T is the smallest triangle that touches S.

Problem 18, page 20

To begin the problem check that, when a and b are integers then the rectangle can be easily subdivided into ab equal sized squares, all of side length 1.

Suppose a or b is not an integer and a/b can be expressed as a fraction m/n, where m and n are positive integers. Then $a = cm$ and $b = cn$ for some number c. Thus take small squares of side length c and there are certainly mn such squares fitting inside the rectangle.

If a/b is not a fraction (i.e., it is an irrational number) then there would be no choice of side length c for the small squares to work out. In modern language two real numbers a and b are *commensurable* if a/b is a rational number (i.e., a fraction). Thus the answer to the problem is that we must require a and b to be commensurable.

Problem 19, page 20

We just saw in Problem 18 that a rectangle R cannot be tiled with equal squares unless the sides of the rectangle are commensurable. It is also true for any tiling by a collection of squares that this same condition must be met. A proof that a rectangle can be so tiled if and only if a and b are commensurable is given in

> R. L. Brooks, C. A. B. Smith, A. H. Stone and W.T. Tutte, *The dissection of rectangles into squares,* Duke Math. J. (1940) 7 (1): 312–340.

Probably the first proof of the theorem that *a rectangle can be squared if and only if its sides are commensurable* is by Max Dehn,

> Max Dehn, Über Zerlegung von Rechtecken in Rechtecke, Mathematische Annalen, Volume 57, September 1903.

though it might be rather more inaccessible to most of our readers.

Chapter 2

Pick's Rule

Look at the polygon in Figure 2.1. How long do you think it would take you to calculate the area? One of us got it in 41 seconds. No computers, no fancy calculations, no advanced math, just truly simple arithmetic. How is this possible?

The projects in this chapter have as their centerpiece work published in 1899 by Georg A. Pick (1859–1942). His theorem supplies a remarkable and simple solution to a problem in areas. Set up a square grid with the dots equally spaced one inch apart and draw a polygon by connecting some of the dots with straight lines. What is the area of the region inside the polygon?

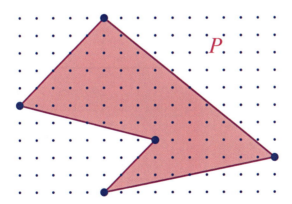

Figure 2.1: What is the area of the region inside the polygon?

You will likely imagine counting up the number of one-inch squares inside and then making some estimate for the partial squares near the outside. Pick's Rule says that the area can be computed *exactly* and *quickly*: look at the dots!

As is always the case in this book, it is the *discovery* that is our main goal. Many mathematics students will learn this theorem in the traditional way: the theorem is presented, a few computations are checked, and the short inductive proof is presented in class. We take our time to try to find out how Pick's formula might have been discovered, why it works, and how to come up with a method of proof.

2.1 Polygons

In Figure 2.1 we have constructed a square grid and placed a polygon on that grid in such a way that each vertex is a grid point. The main problem we address in this chapter is that of determining the area inside such a polygon. We need to clarify our language a bit, although the reader will certainly have a good intuitive idea already as to what all this means.

Familiar objects such as triangles, rectangles, and quadrilaterals are examples. Since we work always on a square grid the line segments that form the edges of these objects must join two dots in the grid.

2.1.1 On the grid

We can use graph paper or even just a crude sketch to visualize the grid. Formally a mathematician would prefer to call the grid a *lattice* and insist that it can be described by points in the plane with integer coordinates[1].

But we shall simply call it *the grid*. It will often be useful, however, to describe points that are on the grid by specifying the coordinates.

Problem 20 *A point (m,n) on the grid is said to be visible from the origin $(0,0)$ if the line segment joining (m,n) and $(0,0)$ contains no other grid point. Experiment with various choices of points that are or are not visible from the origin. What can you conclude?*   Answer □

2.1.2 Polygons

It is obvious what we must mean by a triangle with its vertices on the grid. Is it also obvious what we must mean by a polygon with its vertices on the grid? We certainly mean that there are n points

$$V_1, \ V_2, \ V_3, \ \ldots, \ V_n$$

on the grid and there are n straight line segments

$$V_1V_2, \ V_2V_3, \ V_3V_4, \ \ldots, \ V_nV_1 \qquad (n \geq 3)$$

joining these pairs of vertices that make up the edges of the polygon. Figure 2.2 illustrates. Need we say more?

Problem 21 *Consider some examples of polygons and make a determination*

---

[1]It is usual for mathematicians to describe the integers

$$\ldots, -4, \ -3, \ -2, \ -1, \ 0, \ 1, \ 2, \ 3, \ 4, \ldots$$

by the symbol $\mathbb{Z}$ (the choice of letter Z here is for Zahlen, which is German for "numbers"). Then the preferred notation for the grid consisting of all pairs (m,n) where m and n are integers (positive, negative, or zero) would be $\mathbb{Z}^2$.

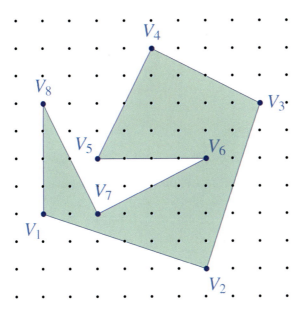

Figure 2.2: A polygon on the grid.

as to whether the statement above adequately describes a general polygon on the grid.   Answer □

2.1.3 Inside and outside

A polygon P in the plane divides the plane into two regions, an inside and an outside. Points inside of P can be joined by a curve that stays inside, while points outside can be joined by a curve that stays outside. If you travel in a straight line from a point inside to a point outside then you will have crossed the polygon. All these facts may seem quite obvious, but a proof is not easy.

Nor is it as obvious as simple pictures appear to suggest. Imagine a polygon with thousands of vertices shaped much like a maze or labyrinth. Take a point somewhere deep in the maze and try to decide whether you are inside or outside of the polygon. We might be convinced that there is an inside and there is an outside but it need not be obvious which is which.

For these reasons we merely state this as a formal assumption for our theory:

2.1.1 *Every polygon P in the plane divides the plane into two regions, the* inside *of P and the* outside *of P. Any two points inside (outside) of P can be joined by a curve lying inside (outside) P. But if a line segment has one endpoint inside P and the other outside P, then this line segment must intersect P.*

It is common to call the inside a *polygonal region*, to refer to the polygon itself as the *boundary* of the polygonal region, and to refer to points inside but not on the boundary as *interior points*. For simplicity, we often refer simply to the *inside* of the polygon.

Problem 22 *If you are given the coordinates for the vertices of a polygon spec-ified in order and the coordinates of some point that is not on the polygon, how might you determine whether your point is inside or outside the polygon?*

Answer ☐

2.1.4   Splitting a polygon

A polygon *can be split* into two smaller polygons if there is a line segment *L* joining two of the vertices that is inside the polygon and does not intersect any edge of the polygon (except at the two vertices which it joins). Figure 2.3

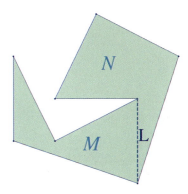

Figure 2.3: Finding a line segment *L* that splits the polygon.

illustrates one particular case. The large polygon with eight vertices has been split into two polygons *M* and *N*. The polygon *M* has five vertices and the polygon *N* also has five vertices.

This splitting property is fundamental to our ability to prove things about polygons. If every polygon can be split into smaller polygons we can prove things about small polygons and use that fact to determine properties that would hold for larger polygons.

Problem 23 *Figure 2.3 shows one choice of line segment L that splits the poly-gon. How many other choices of a line segment would do the split of the large polygon?*   Answer ☐

Problem 24 *Experiment with different choices of polygons and determine which can be split and which cannot. Make a conjecture.*   Answer ☐

Problem 25 *Prove that, for every polygon with four or more vertices, there is a pair of vertices that can be chosen so that the line segment joining them is inside the polygon, thus splitting the original polygon into two smaller polygons.*

Answer ☐

Problem 26 *In Figure 2.3 the large polygon has eight vertices. It splits into two polygons M and N each of which has five vertices. Each of the smaller polygons has fewer vertices than the original eight. Is this true in general?*   Answer ☐

2.1.5 Area of a polygonal region

A polygonal region (the inside of a polygon) has an *area*. This is rather more straightforward than the statement about insides and outsides. If you can accept the elementary geometry that you have learned (the area of a rectangle is given by length × width, the area of a triangle is given by 1/2 × base × height) then polygonal area is simple to conceive. Break the polygon up into small triangles (as in Figure 2.4 for example); then the area would be simply the sum of the area of the triangles. Figure 2.4 is considered a *triangulation* of Figure 2.1.

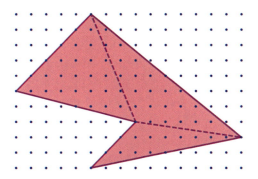

Figure 2.4: A triangulation of the polygon in Figure 2.1.

There are more sophisticated theories of area but we don't need them for our process of discovery here. It is really quite clear in any particular example how to triangulate and therefore how to find the area. Better is to show that any polygon can be triangulated.

Problem 27 *Figure 2.4 illustrates a triangulation of the polygon P. Can you find a different triangulation?* Answer □

Problem 28 *Using the splitting argument of Section 2.1.4 show that every polygon can be triangulated by joining appropriate pairs of vertices.*

Answer □

2.1.6 Area of a triangle

Let begin with an elementary geometry problem. We ask for the area of a triangle with its three vertices at the points $(0,0)$, (s,t), and (a,b) on the grid. Figure 2.5 illustrates one possible position for such a triangle. This problem will not necessarily help solve our main problem (finding a simple method for all polygons) but it will be an essential first step in thinking about that problem.

What method to use? The first formula for the area of a triangle that all of us learned is the familiar

$$1/2 \times \text{base} \times \text{height}.$$

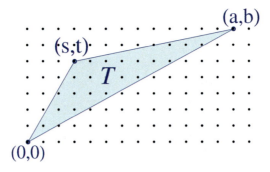

Figure 2.5: Triangle with one vertex at the origin.

With that formula can we easily find the area of all triangles on the grid? Yes and no. Yes, we can do this. No, sometimes we wouldn't want to do it this way.

We can find (although not without some work) the length of any side of a triangle since the corners are at grid points. But finding the height would not be so obvious unless one of the sides is horizontal or vertical.

Is there a formula for the area of a triangle knowing just the lengths of the three sides. Should we pursue this?

Seem reasonable? Given a triangle on the grid we can use the Pythagorean theorem to compute all the sides of the triangle. Once you know the sides of a triangle you know exactly what the triangle is and you should be able to determine its area.

Heron's formula Search around a bit (e.g., on Wikipedia) and you will likely find Heron's formula. If a triangle T has side lengths a, b, and c then
$$\text{Area}(T) = \sqrt{s(s-a)(s-b)(s-c)}$$
where
$$s = \frac{a+b+c}{2}$$
is called the semiperimeter of T (since it is exactly half of the triangle's perimeter). Wikipedia lists three equivalent ways of writing Heron's formula:
$$\text{Area}(T) = \frac{1}{4}\sqrt{(a^2+b^2+c^2)^2 - 2(a^4+b^4+c^4)}$$
$$\text{Area}(T) = \frac{1}{4}\sqrt{2(a^2b^2+a^2c^2+b^2c^2) - (a^4+b^4+c^4)}$$
and
$$\text{Area}(T) = \frac{1}{4}\sqrt{(a+b-c)(a-b+c)(-a+b+c)(a+b+c)}.$$

While all this is true and we could compute areas this way, it doesn't appear likely to give us any insight. Well, these computations will work, but after a long series of tedious calculations we will not be any closer to seeing how to find easier ways.

So, in short, not a bad idea really, just one that doesn't prove useful to our problem. This problem should encourage you to find a different way of computing the area of triangles on the grid.

Decomposition method to compute triangle areas A better and easier method for our problem is to decompose a larger, easier triangle that contains this triangle. Then, since the pieces must add up to the area of the big triangle (which we can easily find) we can figure out the area of our triangle by subtraction.

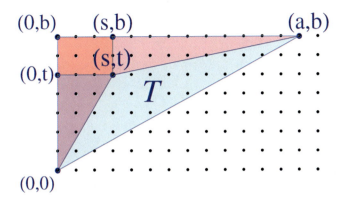

Figure 2.6: Decomposition for the triangle in Figure 2.5

In Figure 2.6 we show a larger triangle containing T that has vertices at $(0,0)$, $(0,b)$, and (a,b). This triangle has base a and height b and so area $ab/2$. The figure shows the situation for the point (s,t) lying above the line joining the origin and (a,b) and $t < b$. There are other cases. Problem 29 asks you to verify that the formula we obtain is valid in all cases.

In the figure we see, in addition to T itself, two triangles and a rectangle. The dimensions of the rectangle are s by $b - t$. The base and height of the triangle below the rectangle are s and t; the dimensions of the triangle to the right of the rectangle are $b - t$ by $a - s$. Thus this decomposition of the large triangle must give

$$\frac{ab}{2} = \text{Area}(T) + s(b-t) + \frac{st}{2} + \frac{(a-s)(b-t)}{2}.$$

The rest is now algebra, but fairly simple if a bit longer than you might prefer. We see that

$$\text{Area}(T) = \frac{1}{2}\{ab - 2(s(b-t) - st - (a-s)(b-t)\}$$

Tidy this up and find that

$$\text{Area}(T) = \frac{at - bs}{2}.$$

You should be able to verify that, in the cases we didn't consider for the location of the point (s,t), we obtain the same formula, or the formula with the sign reversed, that is

$$\text{Area}(T) = \frac{bs - at}{2}.$$

The simplest way to report our findings is to give the formula

$$\text{Area}(T) = \left| \frac{at - bs}{2} \right|$$

which is valid in all cases. (This is Problem 29.)

This is likely more algebra that most of our readers would care to see. Nothing here was all that difficult however. This formula is not simple enough to be a candidate for our "simple" area calculation formula.

Problem 29 *Figure 2.6 shows how to compute the area of a triangle T that has vertices at $(0,0)$, (s,t), and (a,b) but only in the special case shown for which (s,t) lies above the line joining $(0,0)$ and (a,b) with $t < b$. Draw pictures that illustrate the remaining positions possible for the point (s,t) and show that in each of these cases the formula*

$$\text{Area}(T) = \left| \frac{at - bs}{2} \right|$$

is valid.   □

Problem 30 (Area experiment) *Try computing a number of areas of polygons with vertices on the grids, record your results and make some observations.*

Answer □

Problem 31 *Show that the area of every triangle on the grid is an integer multiple of $1/2$.*

Answer □

Problem 32 *Use Problem 31 to show that the area of every polygon on the grid is an integer multiple of $1/2$.*   Answer □

2.2   Some methods of calculating areas

Before attacking our area problem let us take a short digression to consider some possible methods of computing areas. How long do you think it would take to calculate the area inside the polygon P of Figure 2.7 that started this chapter by any of the methods we have so far discussed?

The method we have already suggested for doing the computation would require us to break up P into the three triangles displayed in Figure 2.4, compute the area of each, and then add up the three areas. But you would notice that none of the three triangles has a horizontal or vertical side. It would take

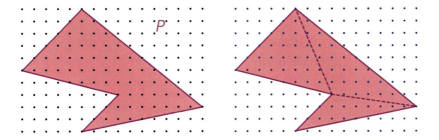

Figure 2.7: The polygon P and its triangulation

some calculating to determine the areas of these triangles. The methods of Section 2.1.6 would certainly work for each of these three triangles and so, in a reasonable amount of time, we could indeed compute the area of the polygon.

This is not impressive, however, and takes far longer than the 41 seconds that we claimed in our introduction. We should consider some other approaches.

2.2.1   An ancient Greek method

Let's look at another method that dates back to the ancient Greeks. They devised a method[2] for approximating the area of any shaped region.

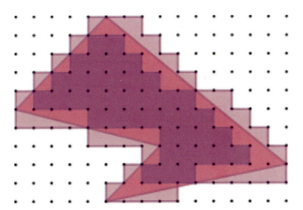

Figure 2.8: Too big and too small approximations

Figure 2.8 shows the polygon with some grid squares highlighted. If we count the grid squares that lie entirely inside P, and add up their areas, we have an approximation to the area inside P. This approximation is too small, because we have not counted the contributions of the squares that lie only partially inside P.

We could also obtain an approximation to the area that is too large by including the full areas of those squares that lie partially inside and partially outside P. The exact area is somewhere between these two approximations. If we do

---

[2]The ancient Greeks would not have used this method for finding areas of polygons. It would be used for circles and other figures that couldn't be broken into triangles.

this for the polygon in Figure 2.7, we find the two approximations are not that close to each other. This is so because there are so many grid squares, each of area 1, that are only partially inside P. the difference between counting them and not counting them is relatively large.

The method of exhaustion   The two approximations will improve if we used smaller grid squares. They would improve again if we used even smaller grid squares.

Suppose each grid square were subdivided into 4 smaller squares and the process were repeated. Do you see that the excess of counting the partial squares is reduced, while the approximation obtained by not counting them is increased. In a more advanced course one could show that by using smaller and smaller squares, one can obtain the exact area using the theory of limits. The approximations that are too small increase towards the area, while the approximations that are too big decrease towards the actual area.

> How long do you think it would take to find the area of P using this method?

This method is sometimes called the *method of exhaustion* which refers to the fact that the area is exhausted by each step although, as you can well imagine, it might be the person doing the computations that is exhausted.

One wouldn't actually have to compute *all* those approximating areas. A person well-versed with the limit process could obtain formulas for the approximating areas at an arbitrary stage of the subdividing process and could then calculate the limit. Still—not a quick process, probably slower than calculating the area by our first method.

2.2.2   Grid point credit—a new fast method?

Now for our purposes, the sizes of our squares are fixed – they all have area 1. To get an exact area we would have to calculate the exact areas of the parts of the partial squares that lie inside P.

Is there a connection between the number of grid points and the number of grid squares inside a grid polygon? Perhaps we can find a way of assigning "grid point credit" to grid points that mimics the approximations we discussed. Since we don't have the option of reducing the size of grid squares, we seek a formula that gives an exact area, not one that requires some sort of limit. Perhaps we can do this by giving credit to points depending on their location inside the polygon. Let's see if we can formulate a method of assigning full or partial credit to grid points.

If we were dealing with the whole plane, rather than with the inside of a polygon, we would note that every grid point is a corner point of four squares, and every grid square had four grid points as corners. Thus one could count grid squares by counting grid points. Of course, we are not dealing with the whole plane, we are dealing with a polygon. But it does suggest a start.

Assigning credit   When a grid point p is "well inside" the polygon, all four squares that have p as a corner are inside P. Let's try giving full credit of 1 to such points.

What about other points? When only a certain part of the four squares that have the point as a corner lies inside P, we try giving that point proportional credit.

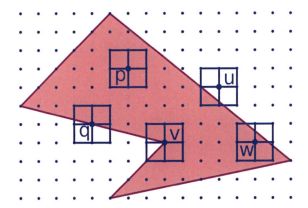

Figure 2.9: Polygon P with 5 special points and their associated squares

Notice there are several grid points, such as the point q, on an edge of P, many grid points like p "well inside" P, points like w that are inside P but near an edge, vertices like v and points like u that are outside of P but near an edge.

In this simple figure, we see that only half of the area of the four squares that have q as a corner lies inside P. Let's try half credit for q. You can check that the same is true of all grid points that are on an edge of P, except the vertices where a similar picture would suggest credit different from 1/2.

We have already determined that the point p deserves credit equal to 1 because the four squares associated with p lie inside P.

At w the 4 associated squares appear to be more than half filled with points of P, so w should get more than 1/2 credit. The vertex v should receive more than 1/2 credit. Even points like u that are outside but near P deserve some credit. The exact amount of credit each of these grid points deserves has to be calculated.

We can do this type of calculation for all grid points inside, on, or near P, add up all the credits and get the exact area of P.

Is this useful or practical?   This would be useful if there were a way of assigning credit to grid points in a simple way, based only on their location. Points well inside P, like p would get full credit, and all other points whose associated squares contain points inside P (like q, w, v and u) would get credit between 0 and 1, based on the percentage of the area of the four associated squares that lies inside P.

Will adding up all these credits give us the exact area? Yes, it will.

Is this practical? Is it easy? Would all grid points on an edge of a polygon (except vertices) deserve credit exactly 1/2? Look at Figure 2.10.

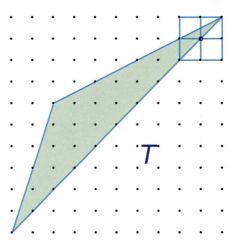

Figure 2.10: A "skinny" triangle.

Here the point p is located on the boundary of the triangle T at (9,9). Our earlier example suggested that such a boundary point should receive credit 1/2. But less than half of the area of the four squares having the point p as a corner lies inside T. So this point p doesn't deserve half credit after all: it deserve less. We'd have to do a calculation to determine the credit this point deserves, even though it lies on an edge of T. That would defeat our purpose of finding a simple and quick method of obtaining the area.

We see that just knowing the location of a point gives no immediate clue as to the proper credit, unless the point is well inside the polygon, or well outside it. A possibly messy calculation would be necessary to determine its proper credit.

> How long do you think it would take to find the area of P using this method?

The answer is "Way too long." The process would involve so much calculation that for practical purposes it is useless.

Some other kind of credit?   What now? We can give up the idea of assigning grid points credit. Or, we can keep that idea, but use what we have learned from our earlier experiments to find a way that does lead to a simple, practical method of calculating the area.

This sort of situation often occurs in mathematical discovery. A plausible approach looks promising at first, but does not achieve the desired outcome. Instead of giving up, the researcher retains part of that approach, but makes use of earlier experimentation and earlier results to find a similar method that has the desired outcome. In this case, it involves discovering the correct simple and quick way to assign credit to grid points.

2.3 Pick credit

The grid point credit idea based on area works certainly. It is entirely general since it offers a method to compute the area of any figure. The figure need not be a polygon nor need it have any points on the grid itself for this to work. The method assigns a value between 0 and 1 for every grid point but the nature of the point offers no help in guessing at the credit—it must be computed in each case. The only exception is that points well-inside the polygon clearly get a grid point credit of 1 and points well-outside get a zero credit.

Because the method is so general we do not expect it to offer much insight into the current problem. Nor is this method easy or fast. We want a fast and easy method for computing polygonal areas and we want a method that explains transparently why the areas are invariably multiples of $1/2$ (as we saw in Problem 32).

We will still use the idea of assigning a value to each grid point but, encouraged by our earlier experiments and observations, we will assign only values of 0, $1/2$, or 1. We will not attempt to assign values that imitate the grid point credit values. Points with a small grid point credit might well require us to assign 1 or $1/2$ and points with a large area assignment might well require us to assign 0 or $1/2$.

We can call this *Pick credit* with the understanding that it will be in almost no way related to the grid point credit method we have just proposed. As we have seen in working with grid point credit, the credit each point gets simply must be computed: there is no way of looking at a point and deciding that some feature of the point justifies more or less credit.

For the Pick count we want to do no computations, although we are willing to look for any features of the point that might require different credits. We cannot decide whether a point *deserves* credit (in the same way that the area credit computations did). We must simply experiment with different possible assignments until we find the one that works.

2.3.1 Experimentation and trial-and-error

In order to get some familiarity with our problem let us compute some areas for a variety of polygonal regions constructed on grids. These problems are essential training for our task and help reveal the true nature of the problem we are trying to solve. One goal we have, in addition just to familiarization with area problems, is that of finding the appropriate Pick credit that might work for our area problem.

A good starting point is to investigate the area of primitive triangles. A triangle on the grid must have all three vertices on the grid. If it contains no other grid points then it is called a *primitive triangle*.

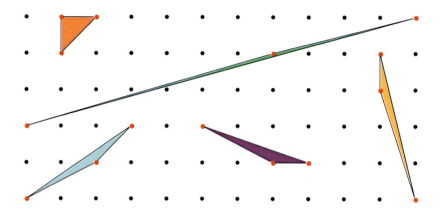

Figure 2.11: Some primitive triangles.

Problem 33 (Primitive triangles) *What can you report about the area of primitive triangles?*

Answer □

Problem 34 *Find a number of triangles that have vertices on the grid and contain only one other grid point, which is on the edges of the triangle. What did you observe for the areas?*   Answer □

Problem 35 *Find a number of triangles that have vertices on the grid and contain only one other grid point, which is inside the edges of the triangle. What did you observe for the areas?*   Answer □

Problem 36 *In Figure 2.12 we see a collection of four polygons each of which has 4 boundary points and 6 interior points. Compute the areas and comment.*

Answer □

Problem 37 *Show that it is possible to construct a polygon on the grid that has as its area any one of the numbers*

$$\frac{1}{2}, \ 1, \ \frac{3}{2}, \ 2, \frac{5}{2}, \ 3, \ \frac{7}{2}, \ 4, \ldots$$

Answer □

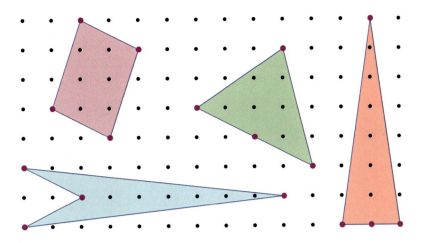

Figure 2.12: Polygons with 4 boundary points and 6 interior points

Problem 38 *What numbers can appear as the area of a square on the grid? Experiment with various possibilities and then explain the pattern you see.*

Answer □

Problem 39 *Look at Figure 2.13. Compute the area of the rectangle R and the triangle T. Try assigning a Pick credit of 1 to every point that is inside P and a Pick credit of 0 to every point that is not. Points on the boundary or outside get 0 credit. Consider how the area of interest compares with the total of the credits. Try some other simple figures as well.*

Answer □

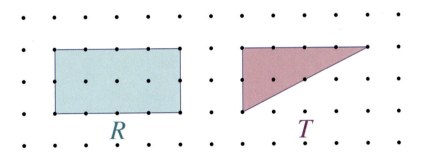

Figure 2.13: Compute areas.

Problem 40 *Repeat the preceding exercise but this time try assigning credit of 1 to every point that is inside or on P. Points outside get zero credit.*

Answer □

Problem 41 *Can you see a way to improve the approximation in Problem 40 by giving less credit for the grid points that lie on the polygon (i.e., on the edges of the rectangle R or of the triangle T)?*

Answer □

Problem 42 *Repeat Problem 41 with a few more examples using rectangles and triangles with corners on the grid. (It simplifies the computation if you choose rectangles with horizontal and vertical sides and triangles with one vertical side and one horizontal side.)*   Answer □

2.3.2   Rectangles and triangles

Our exploration in Problems 39–42 has suggested a first estimate of the form

$$\text{Area}(P) \approx [\text{\# of grid points inside } P] + \frac{[\text{\# of grid points on } P]}{2}$$

using our idea of full credit for the inside points and half-credit for the boundary points. We cannot say that the area is *equal* so we are using here the symbol $\approx$ to suggest that this is an approximation or a crude first estimate.

If we use I to denote the count for the interior grid points and B for the count of the boundary grid points then a *Pick count*

$$I + \frac{B}{2}$$

gets close to the areas that we have considered so far.

Example 2.3.1 Here is another computation that suggests that half-credit is exactly right for the assignment of credit to the boundary grid points. The rectangle in Figure 2.13 can be split into two triangles as shown in Figure 2.14.

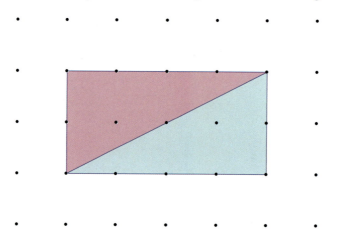

Figure 2.14: Split the rectangle into two triangles.

There is one interior point inside the rectangle that becomes a boundary point for the two triangles. In the estimate for the rectangle that interior point gets full credit. For the triangles it has become a boundary point, and so gives only half-credit to each of the triangles. This is appropriate since the area of each of the triangles is exactly half the area of the rectangle.

The rectangle has 3 interior grid points, 12 grid points on the boundary, and area 8. Each triangle has 1 interior point, 8 points on the boundary, and area 4.

So, as we found in the problems, the formula above gives a first estimate of 9 for the area of the rectangle and 5 for each of the rectangles. In both cases this is 1 more than the correct values. ◄

Problem 43 *Determine an exact formula for the area of rectangles with vertical and horizontal sides and with vertices on the grid work. Compare with the actual area.* Answer □

Problem 44 *Determine an exact formula for the area of triangles with one vertical side and one horizontal side and with vertices on the grid work. Compare with the actual area.* Answer □

2.3.3 Additivity

One of the key properties of area is *additivity*. If two triangles, two rectangles, or any two polygons that have no common interior points are added together the resulting figure has an area that is equal to the sum of its parts. Certainly then, Pick's formula, if it is a correct way to compute area, must be additive too in some way.

Let us introduce some notation that will help our thinking. For any polygon P we simply count the points in or on the polygon, assigning credit of 1 for points inside and $1/2$ for points on the polygon. Call this *Pick's count* and write it as

$$\text{Pick}(P) = I + \frac{B}{2}.$$

The value I simply counts interior points and B counts boundary points. We are nearly convinced, at this stage, that Pick's count does give a value that is 1 more than the area. Is Pick's count additive?

Suppose M and N are polygons with a common side L but no other points inside or on the boundaries in common. Then M and N can be added to give a larger polygon with a larger area as in Figure 2.15. Call it P. The larger polygon has all the edges of M and N except for L which is now inside the large polygon P.

Now we wish to show that we can determine $\text{Pick}(P)$ from

$$\text{Pick}(M) + \text{Pick}(N).$$

Then we want to use this fact to advantage in our computations.

Problem 45 *We know that*

$$\text{Area}(P) = \text{Area}(M) + \text{Area}(N).$$

Compare $\text{Pick}(P)$ *and* $\text{Pick}(M) + \text{Pick}(N)$. *In fact, show that*

$$\text{Pick}(M) + \text{Pick}(N) = \text{Pick}(P) + 1.$$

Answer □

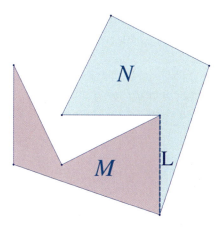

Figure 2.15: Adding together two polygonal regions.

Problem 46 *Write a simpler and more elegant solution of Problem 44 using the notation of this section.*   Answer □

Problem 47 *Suppose that a polygon P has been split into three smaller polygons P_1, P_2, and P_3 by adding two lines joining vertices. Show that*

$$\text{Pick}(P_1) + \text{Pick}(P_2) + \text{Pick}(P_3) = \text{Pick}(P) + 2.$$

□

2.4   Pick's formula

We have established the formula

$$\text{Area}(P) = \text{Pick}(P) - 1 = I + \frac{B}{2} - 1$$

for certain rectangles and for certain triangles. Any polygon which we can break up into parts comprised of such rectangles and such triangles can then be handled by the additivity of areas and the additive formula for the Pick count using methods we have already illustrated. If you think of some more complicated polygons, you might find that they can be broken up into triangles, but not necessarily triangles with one horizontal side and one vertical side.

Let's first experiment with a particular example of a triangle that does not meet those requirements.

Example 2.4.1 Let's try our formula on the triangle in Figure 2.16. The base of this triangle has length 10 and its altitude is 8. Thus its area is 40.

Our conjectured formula uses 33 interior points and 16 boundary points, giving an answer of

$$33 + 16/2 - 1 = 40$$

for the area, which is the same answer.   ◄

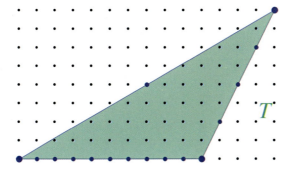

Figure 2.16: A triangle with a horizontal base.

The formula works but we have not seen why since we merely did a computation. We might try to check that this formula would work for all triangles with a horizontal base (this is Problem 48). Then we could try a more ambitious problem and determine that all triangles have the same property (this is Problem 49). Problem 48 is just a warm-up to the full case and is not needed. Problem 49 can be proved just by knowing that this formula is correct for rectangles and for triangles with both a horizontal and a vertical side.

Problem 48 *Show that the area of any triangle T with vertices on the grid and with a horizontal base is given by the formula*

$$\text{Area}(T) = \text{Pick}(T) - 1.$$

Answer □

Problem 49 *Show that the area of any triangle T (in any orientation) with vertices on the grid is given by the formula*

$$\text{Area}(T) = \text{Pick}(T) - 1.$$

Answer □

2.4.1 Triangles solved

The figures that we saw in the answer for Problem 49, duplicated here as Figure 2.17, are the most complicated ones that can arise if one wishes to follow the method suggested. The key idea is that triangles in any odd orientation can be analyzed by looking at rectangles and triangles in a simpler orientation. It is the additivity properties of areas and of Pick counts that provides the easy solution.

Let us revisit Problem 49 and provide a clear and leisurely proof. We need to analyze the situation depicted in the right-hand picture in Figure 2.17. Here we have labeled the first triangle as T_0: this is the triangle in a strange orientation for which we do not yet know that the Pick rule will work.

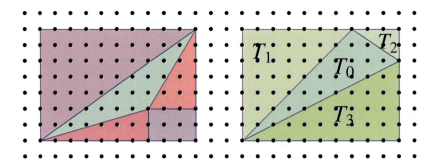

Figure 2.17: Triangles in general position.

The remaining triangles T_1, T_2, and T_3 are all in a familiar orientation and we can use Pick's rule on each of them. Together they fit into a rectangle R for which, again, we know Pick's rule works.

The additivity of areas requires that

$$\text{Area}(R) = \text{Area}(T_0) + \text{Area}(T_1) + \text{Area}(T_2) + \text{Area}(T_3).$$

The additivity rule for the Pick count we have seen in the previous section:

$$\text{Pick}(R) + 3 = \text{Pick}(T_0) + \text{Pick}(T_1) + \text{Pick}(T_2) + \text{Pick}(T_3).$$

The extra 3, we remember, comes from the fact that three pairs of vertices are recounted when we do the sum.

Now we just have to put this together to obtain the formula we want, namely that

$$\text{Area}(T_0) = \text{Pick}(T_0) - 1.$$

Problem 50 *Do the algebra to check that*

$$\text{Area}(T_0) = \text{Pick}(T_0) - 1.$$

Answer □

Problem 51 *Consider once again the polygon P in Figure 2.1. What would Pick's formula give for the area of the P? Triangulate the polygon, use Pick's formula for each triangle, add up the areas, and compare with the area that you just found.*

Answer □

2.4.2 Proving Pick's formula in general

We have so far verified that the formula works for triangles in any orientation. We should be ready now for the final stage of the argument which uses the triangle case to start off an induction proof[3] that solves the general case.

---

[3] See the Appendix for an explanation of mathematical induction if you are not yet sufficiently familiar with that form of proof.

The key stage in your induction proof will have to use the *splitting argument* that we saw in Section 2.1.4. Use mathematical induction on the number of sides of P and, at the critical moment in your proof, use the splitting argument to reduce a complicated polygon to two simpler ones.

Problem 52 *Prove that the formula*

$$\text{Area}(P) = \text{Pick}(P) - 1$$

works for every polygon P having vertices on the grid. Answer □

2.5 Summary

We have obtained a quick, easy, accurate formula for calculating the area inside any polygon having vertices on the grid.

Try this formula on the polygon in Figure 2.18 where we have made the task of spotting the appropriate grid points somewhat easier. How long did it take? Did you improve the record of 41 seconds?

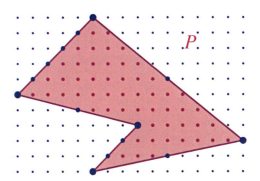

Figure 2.18: Polygon P with border and interior points highlighted.

Let's review our method of discovery. In Section 2.1.6 we revisited some formulas for the area of a triangle that we might have learned in elementary geometry. These formulas did give the area of a triangle, but would often involve some unpleasant computations. (We were seeking something quick and easy.) We were able to use such formulas to prove that every polygon with vertices on the grid has an area that is an integral multiple of 1/2. (Problem 32)

We proceeded in Section 2.2 to discuss some other methods for computing areas of polygons. None of these met our requirement of quickness and ease of computation. One of these suggested a notion of giving "credit" to grid points inside, on, or near the polygon. To calculate an area by this method would often involve a huge amount of messy calculation, so it was an impractical method. But it did suggest a method of giving credit to grid points. Our experiences in solving the problems of Section 2.3.1 suggested that only 0, 1/2 and 1 should be considered as possible credits.

So we did some more experimentation, in Section 2.3, based on our observations. We did some calculations for relatively simple polygons, and arrived
at a formula that actually gave the area for a variety of cases – in particular for
rectangles and triangles that have at least two sides that are vertical or horizontal. By now it became natural to suspect that the formula we obtained would
actually apply to all polygons with vertices on the grid. But we had some more
checking to do – we hadn't yet checked more complicated polygons, even triangles whose sides are not vertical or horizontal.

In Section 2.4 we put it all together. First, we established the result for all
triangles with vertices on the grid, regardless of their orientation. Then we used
mathematical induction to verify the formula for all polygons with vertices on
the grid. We had accomplished our goal.

The role of induction   By the time we came to the actual proof by induction,
we were (almost) convinced that the formula is correct. The discovery part was
complete. We used induction only for verification purposes. It was not part of
the discovery process.

This will be true of every use of mathematical induction in this book. By
the time we get to the induction step, we are almost convinced that the result we
obtained is correct. The induction step removes all doubts.

Other methods   There are many other approaches to proving Pick's formula.
Some of the material in Sections 2.6.3, 2.6.5. 2.6.7, and 2.6.8 discuss other
approaches that shed some additional light on the subject. In a later chapter in
Volume 2 we will use some graph theory and a theorem of Euler to revisit Pick's
theorem.

2.6   Supplementary material

2.6.1   A bit of historical background

A bit more historical detail on Pick himself is given in the
article by M. Ram Murty and Nithum Thain ([17] in our
bibliography) from which the following quote is taken:
"Pick was born into a Jewish family in Vienna on August 10,
1859. He received his Ph.D. from the University of Vienna under
the supervision of Leo Koenigsberger in 1880. He spent most of
his working life at the University of Prague, where his colleagues
and students praised his excellence at both research and teaching.
In 1910, Albert Einstein applied to become a professor of theo-
retical physics at the University of Prague. Pick found himself
on the appointments committee and was the driving force in getting Einstein accepted. For the brief period that Einstein was at

Figure 2.19: Pick

Prague, he and Pick were the closest friends. They were both
talented violinists and frequently played together. In 1929, Pick
retired and moved back to his hometown of Vienna. Nine years later, Austria was an-
nexed by Germany. In an attempt to escape the Nazi regime, Pick returned to Prague.
However, on July 13, 1942, he was captured and transported to the Theresienstadt con-
centration camp. He passed away there thirteen days later, at the age of 82.
Pick's formula first came to popular attention in 1969 (seventy
years after Pick published it) in Steinhaus's book *Mathematical
Snapshots*."

Pick's theorem was originally published in 1899 in German (see [7] in our
bibliography). Recent proofs and extensions of Pick's theorem can be found
in several American Math. Monthly articles by W. W. Funkenbusch [4], Dale
Varberg [14], and Branko Grünbaum and G. C. Shephard [5].

2.6.2 Can't be useful though

Is Pick's theorem of any use? Not likely, you might say. Here is a remark
though that might change your mind:

> "Some years ago, the Northwest Mathematics Conference was held in Eu-
> gene, Oregon. To add a bit of local flavor, a forester was included on the
> program, and those who attended his session were introduced to a variety
> of nice examples which illustrated the important role that mathematics
> plays in the forest industry. One of his problems was concerned with
> the calculation of the area inside a polygonal region drawn to scale from
> field data obtained for a stand of timber by a timber cruiser. The stan-
> dard method is to overlay a scale drawing with a transparency on which a
> square dot pattern is printed. Except for a factor dependent on the relative
> sizes of the drawing and the square grid, the area inside the polygon is
> computed by counting all of the dots fully inside the polygon, and then
> adding half of the number of dots which fall on the bounding edges of
> the polygon. Although the speaker was not aware that he was essentially
> using Pick's formula, I was delighted to see that one of my favorite math-
> ematical results was not only beautiful, but even useful."

The quote is due to Duane W. Detemple and is cited in the article by Branko
Grünbaum and G.C. Shephard [5].

2.6.3 Primitive triangulations

Primitive triangles play a key role in our investigations. These are the triangles
that contain no other grid points than their three vertices. We saw that each
primitive triangle had area $1/2$ and Pick's formula confirms this.

A *primitive triangulation* of a polygon on the grid is a triangulation with
the requirement that each triangle that appears must be primitive. Figure 2.20

illustrates a polygon that contains two interior grid points leading to a primitive triangulation containing eight primitive triangles.

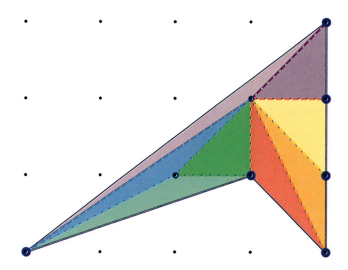

Figure 2.20: A primitive triangulation of a polygon.

How would one go about constructing such a triangulation? Must one always exist? What other features are there?

The splitting game   To study these questions let us introduce a simple *splitting game* that can be played on polygons. Mathematicians frequently introduce games to assist in the analysis of certain problems. We will return to the investigation of games in other chapters.

Two players agree to start with a polygon on the grid and, each taking turns, to split it into smaller subpolygons on the grid. Player A starts with the original polygon and splits it into two (by adding one or two line segments according to rules given below). Player B now faces two polygons. She chooses one of them and splits it into two (by following the same rules). Player A now faces three polygons. He chooses one of them and splits it into two. Player B now faces four polygons. She chooses one of them and splits it into two.

And so on. The game stops when none of the polygons that one sees can be split further. The last person to move is declared the winner.

The rules   The rule for each move is that the player is required to choose a polygon in the figure that has arisen in the play of the game and that has not, as yet, been split. The player then splits that polygon in one of these two ways:

Type 1   The player selects two grid points on the boundary of the polygon. The line segment joining them is constructed provided it is entirely inside the polygon, thus splitting the polygon into two smaller polygons.

Type 2 The player selects two grid points on the boundary of the polygon and also a grid point in the interior. The two line segments joining the interior grid point to the two boundary points are constructed provided they are entirely inside the polygon.

Note that each play of the game splits the original polygon into more and more pieces. More precisely, after the first move the original polygon has been split into two polygons, after the second move there will be three polygons, and after the kth move there will be $k + 1$ polygons. At some point we must run out of grid points that can be joined and the game terminates with a winner declared.

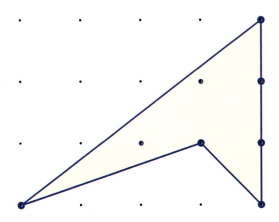

Figure 2.21: A starting position for the game.

Problem 53 *Play the splitting game using the polygon in Figure 2.21 as the starting polygon. What can you report?*   Answer ☐

Problem 54 *Play the splitting game a few times with some simple choices of polygons. What can you report?*   Answer ☐

Problem 55 *Prove that any play of the splitting game always ends with a primitive triangulation of the starting polygon.*

Answer ☐

Problem 56 *Use Pick's formula to compute the area of all primitive triangles.*

Answer ☐

Problem 57 *Suppose that the starting polygon has B grid points on the boundary and I grid points in its interior. Using Pick's theorem, determine how many triangles there are in the final position of the game and how many moves of the splitting game there must be.*   Answer ☐

Problem 58 *Suppose that the starting polygon has B grid points on the boundary and I grid points in its interior. Which player wins the game?*

Answer ☐

2.6.4   Reformulating Pick's theorem

We can reformulate Pick's theorem in terms of primitive triangulations using what we have discovered by playing this splitting game.

We saw that primitive triangulations must exist. We saw that there was always the same number of triangles in any primitive triangulation. We observed that we could count the number of triangles by the formula

$$2I + B - 2.$$

Pick's theorem provided the area of $1/2$ for every primitive triangle. All these facts add up to Pick's theorem and, had we known them, the area formula $I + B/2 - 1$ would have followed immediately. Consequently the following statement is equivalent to Pick's area formula and is a better way of thinking about it and a better way of stating it.

2.6.1 (Pick's Theorem) *A primitive triangulation of any polygon P on the grid exists, and moreover*

1. *The area of any primitive triangle is $1/2$.*

2. *The number of triangles in any primitive triangulation of P is exactly*
$$2I + B - 2$$

where I is the number of grid points inside P and B is the number of grid points on P.

Some people on first learning Pick's area formula ask for an explanation of why such a simple formula works. They see that it does work, they understand the proof, but it somehow eludes them intuitively. But if you ask them instead to explain why the primitive triangulation formula

$$2I + B - 2$$

would work, they see that rather quickly. Of course counting a triangulation of P depends on grid points in and on P. Of course interior points count twice as much as boundary points in constructing a primitive triangulation.

Oddly enough then, thinking too much about areas makes a simple formula more mysterious. Stop thinking about why areas can be explained by grid points and realize that Pick's formula is actually a simple method for counting the triangles in a triangulation. The area formula is merely a consequence of the counting rule for primitive triangulations.

2.6.5   Gaming the proof of Pick's theorem

We used our knowledge of Pick's theorem to analyze completely the splitting game. Not surprisingly, we can use the splitting game itself to analyze completely Pick's theorem.

We know that any splitting game will always result in a primitive triangulation of any starting polygon. We wish to establish that the number of triangles that appear at the end of the game is always given by the formula

$$2I + B - 2$$

where I is the number of grid points inside P and B is the number of grid points on P.

Let us take that formula as a definition of what we mean by *the count*:

$$\text{Count}(P) = 2I + B - 2$$

for any polygon P. Note immediately that if T is a primitive triangle then

$$\text{Count}(T) = 2 \times 0 + 3 - 2 = 1.$$

We play the game on a polygon P splitting it by a Type 1 or 2 move into two polygons M and N. Simply verify that

$$\text{Count}(P) = \text{Count}(M) + \text{Count}(N).$$

This is just a simple counting argument looking at the grid along the splitting line. (Do this as Problem 59).

That means that if we play the game one more step by splitting M into two subpolygons M_1 and M_2 the same thing happens:

$$\text{Count}(M) = \text{Count}(M_1) + \text{Count}(M_2)$$

and so

$$\text{Count}(P) = \text{Count}(M_1) + \text{Count}(M_2) + \text{Count}(N).$$

So, if we play the game to its conclusion, P is split into n primitive triangles T_1, $T_2, \ldots T_n$ in exactly $n - 1$ plays of the game. Consequently

$$\text{Count}(P) = \text{Count}(T_1) + \text{Count}(T_2) + \cdots + \text{Count}(T_n) = 1 + 1 + \cdots + 1 = n.$$

That completes the proof that $\text{Count}(P)$ always gives exactly the correct number of triangles in the primitive triangulation of P.

Problem 59 *We play the game on a polygon P splitting it by a Type 1 or 2 move into two polygons M and N. Verify that*

$$\text{Count}(P) = \text{Count}(M) + \text{Count}(N).$$

Answer □

Problem 60 *Wait a minute! We promised to prove Pick's theorem using the game. We still want to show that for a primitive triangle T,*

$$\text{Area}(T) = 1/2.$$

Can you find a way? [Hint: triangulation works here too.] Answer □

Problem 61 *This proof is simpler, perhaps, than the first proof we gave of Pick's theorem. Why didn't we start with it instead?* Answer □

2.6.6   Polygons with holes

We now allow our polygons to have a few holes. Again we ask for the area of
a polygon constructed on the grid but allowing a hole or perhaps several holes.
The problem itself is not so hard if we can compute the area of the holes since

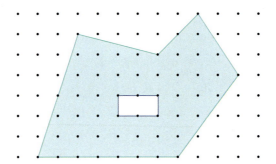

Figure 2.22: What is the area of the polygon with a hole?

then the answer is found by subtracting the area of the holes from the area of
the polygon.

In Figure 2.22 the hole H is a rectangle with area 2; since H is also on the
grid this is easy enough to compute. Indeed if the holes are always polygons
with vertices on the grid we can use Pick's Rule many times to compute all the
areas and then subtract out the holes.

But let us find a more elegant solution. If we use Pick's Rule multiple times
we may end up counting many of the grid points several times. There must be
a simple generalization of the Pick formula

$$\text{Area}(P) = I + B/2 - 1$$

that will accommodate a few holes. Now counting I, we would ignore points
inside the holes. And counting B, we would have to include any boundary points
that are on the edge of the holes.

Polygons with one polygonal hole

Figure 2.23 shows a rectangle P with a hole created by removing a rectangle H
from the inside of P. All of the vertices of R and H are on the grid. Here P
is a 5×12 rectangle and H is a 2×4 rectangle. Thus the area between them
is $60 - 8 = 52$ units. Our objective is to use our counting method directly to
calculate the area between the polygons P and H.

Problem 62 *Experiment with the polygons in Figure 2.23 and others, if necessary, to conjecture a formula for the area between two polygons. As always the
polygons under consideration are to have their vertices on the grid.*

Answer □

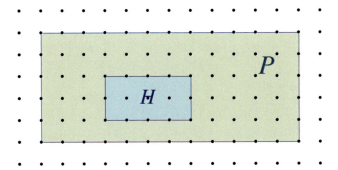

Figure 2.23: Rectangle P with one rectangular hole H.

Problem 63 *Our previous method was (i) counting interior points at full value of 1, (ii) counting points on the boundary of the polygon at half value of 1/2, and finally, (iii) subtracting 1. What goes wrong if we try the same argument for the figure with a hole?* Answer □

Problem 64 (An algebraic argument) *Let us do the entire calculation algebraically. Take P as the outer polygon, H as the hole polygon, and G as the region defined as P take away H. We know from Pick's Rule that*

$$\text{Area}(P) = I(P) + B(P)/2 - 1$$

where by $B(P)$ we mean the number of boundary grid points on P, and by $I(P)$ we mean the number of interior grid points inside P. Similarly

$$\text{Area}(H) = I(H) + B(H)/2 - 1$$

where by $B(H)$ we mean the number of boundary grid points on H, and by $I(H)$ we mean the number of interior grid points inside H.

Find the correct formula for the subtracted area $\text{Area}(P) - \text{Area}(H)$ in terms of $I(G)$ and $B(G)$. Answer □

Polygons with n holes

The algebraic argument we gave is quite general, it applies not only to any polygon P with vertices on the grid and any other such hole polygon H inside P, but also applies (with obvious minor changes) when P has n such polygonal holes inside it. To complete the theory, then try to guess at the final formula and to verify it using the techniques seen so far.

Problem 65 *Determine a formula for the area that remains inside a polygon with n polygonal holes.* Answer □

2.6.7   An improved Pick count

Our Pick-count policy was to assign a count value of 1 for grid points inside the
polygon and a count value of 1/2 for grid points on the polygon itself. This was
certainly successful since it gave us the formula

$$\text{Area}(P) = \text{Pick}(P) - 1$$

which works, as we now have proved, for all possible polygons with vertices at
grid points.

There is another rather compelling and elegant way to do the count. This
makes for a neater proof. This is not a new or different proof, we should point
out. But it is a rather tidy way of expressing the same ideas.

The idea behind it is that the additive formula for the Pick count,

$$\text{Pick}(P) + 1 = \text{Pick}(P_1) + \text{Pick}(P_2)$$

for the situation when the polygon P is split into two polygons P_1 and P_2 with
a common edge is not quite as "additive" as we would prefer: it has this extra
1 that must be included. The additional 1 comes from the two vertices that get
assigned 1/2 in both the counts. That destroys the additivity, but only by a little
bit. To get true additivity we will use the idea that angles are naturally additive.

Angle of visibility   We do a different Pick count. For each point in or on a
polygon P we decide what is its *angle of visibility*. This is the perspective from
which standing at a point we see into the inside of the polygon. For points
interior to P we see a full 360 degrees. For points on an edge but not at a
vertex we see only one side of the edge, so the angle of visibility is 180 degrees.
Finally for points at a vertex the angle of visibility would be the interior angle
and it could be anything between 0 degrees and 360 degrees. We would have to
measure it in each case.

Modified Pick's count   Our *modified Pick's count* is to take each grid point
into consideration, compute its angle of visibility, and divide by 360 to get the
contribution. Points inside get 360/360=1. Points on the edge but not at a
vertex get 180/360=1/2. And, finally, points at the vertex get $a/360$ where a is
the degree measure of the angle. The new Pick count we will write as

$$\text{Pick}^*(P).$$

Add up the count for the vertices   At first sight this seems terribly complicated. How would we be prepared to measure all of the vertex angles? We
would never be able to perform this count. But that is not so.

Take a triangle for example. Except for the three vertices the count is (as
usual) to use 1 for inside points and 1/2 for edge points. The three vertices taken

together then contribute
$$\frac{a+b+c}{360}.$$
While we may have trouble measuring each of angles a, b and c, we know from elementary geometry that the angles in any triangle add up to 180 degrees. So we see that the contribution at the vertices is
$$\frac{a+b+c}{360} = \frac{180}{360} = \frac{1}{2}.$$
The old way of counting would have given us $1/2 + 1/2 + 1/2$ which is 1 larger than this. Thus for any triangle T
$$\text{Pick}^*(T) = \text{Pick}(T) - 1 = \text{Area}(T).$$

In general for a polygon with n vertices it might appear that we would have to compute the angles at each of the vertices to get the contribution
$$\frac{a_1 + a_2 + \cdots + a_n}{360}.$$
But the angles inside any polygon with n vertices add up to $180(n-2)$ degrees. This is because any such polygon can be triangulated in the way we described earlier in the chapter. For example, a quadrilateral can be decomposed into two triangles by introducing a diagonal. Each of the triangles contributes 180 degrees, so the quadrilateral has a total of 2×180 degrees as the sum of its interior angles at the vertices.

Thus we see that the contribution at the vertices of a polygon with n vertices is
$$\frac{a_1 + a_2 + \cdots + a_n}{360} = \frac{180(n-2)}{360} = \frac{n}{2} - 1.$$

Compare the old count to the new count The old way of counting would have given us $1/2$ for each of the n vertices for a total of $n/2$ which is again 1 bigger. Thus we see that for any polygon P
$$\text{Pick}^*(P) = \text{Pick}(P) - 1 = \text{Area}(P).$$
This also explains the mysterious -1 that needed to occur in Pick's formula.

Additivity The ordinary Pick count using $\text{Pick}(P)$ is not quite additive. Every use of the additive rule required a bookkeeping for the addition 1 in the formula
$$\text{Pick}(P) + 1 = \text{Pick}(P_1) + \text{Pick}(P_2).$$
That made our computations a bit messier and gave us a slightly non-intuitive formula
$$\text{Area}(P) = \text{Pick}(P) - 1.$$

Now that we have a better way of counting grid points we have a precisely additive formula
$$\text{Pick}^*(P) = \text{Pick}^*(P_1) + \text{Pick}^*(P_2)$$

and an intuitive area formula

$$\text{Area}(P) = \text{Pick}^*(P).$$

That supplies a different way of writing our proof for Pick's formula that is rather simpler in some of the details. See Problem 67.

Problem 66 *Prove the additive formula*

$$\text{Pick}^*(P) = \text{Pick}^*(P_1) + \text{Pick}^*(P_2)$$

for the modified Pick count for the situation when the polygon P is split into two polygons P_1 and P_2 with a common edge.   Answer □

Problem 67 *Reformulate the proof of Pick's formula using now the modified Pick count to show that*

$$\text{Area}(P) = \text{Pick}^*(P).$$

Answer □

Problem 68 *Determine a formula for the area that remains inside a polygon with n polygonal holes using the modified Pick count idea.*   Answer □

Problem 69 *Does the formula you found in Problem 68 help clarify the formula we have found in Problem 65 for the area inside polygons with holes? Does it explain why that formula needed us to count the number of holes (i.e., why the formula had an n that appeared)?*   Answer □

2.6.8   Random grids

Instead of a square grid let us start off with a large collection of points arranged in any fashion, as for example in Figure 2.24 where the grid points have been chosen at random.

Figure 2.24: Random lattice.

In Figure 2.25 we have constructed a triangle with vertices at grid points of this random lattice. There are three boundary grid points (the three vertices) and two interior grid points; in our usual notation $B = 3$ and $I = 2$. We do not ask for an area computation, but we do ask (as before) whether there must exist a primitive triangulation? We ask too how many triangles would appear in a primitive triangulation of a polygon on this grid?

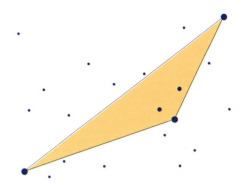

Figure 2.25: Triangle on a random lattice.

Try a few examples until you come to some realization about these problems. The situation is not merely similar to the problem of polygons on square grids: it is identical. In Section 2.6.3 we proved that if P is a polygon on a square grid there must exist a primitive triangulation. In Section 2.6.5 we proved that, if P has I interior grid points and B boundary grid points, then the number of primitive triangles that appear is always exactly $2I + B - 2$. Certainly the same formula works here for the particular case of the triangle in Figure 2.25.

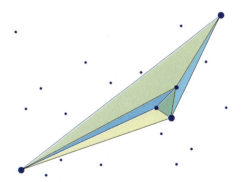

Figure 2.26: Primitive triangulation of the triangle in Figure 2.25.

An examination of our proofs in those sections shows that at no part of the argument did we use any features of a square grid: the points could have been arranged in any fashion at all and the proofs would be unchanged. Hence the result is unchanged: there must always be a primitive triangulation and any such triangulation contains exactly $2I + B - 2$ primitive triangles. The grid points can assume any pattern at all.

When we were concerned about areas then the fact that the grid was square and the points neatly arranged mattered a great deal. When we turn just to counting the pieces of a primitive triangulation the geometry no longer matters. The answer must depend only on the number of boundary points and the number of interior points.

Figure 2.27: Sketch a primitive triangulation of the polygon.

Problem 70 *Sketch a primitive triangulation of the polygon in Figure 2.27 that is on a random grid. How many triangles are there in any primitive triangulation?*

Answer □

2.6.9   Additional problems

We conclude with some additional problems that are related to the material of this chapter.

Problem 71 *Use Pick's Rule to prove that it is impossible to construct an equilateral triangle with its vertices on the dots in a square grid.*

Answer □

Problem 72 (Stomachion) *Find the areas of the polygons in Figure 2.28 by using Pick's Theorem or a simpler method.*

Answer □

Problem 73 *A* Reeve tetrahedron *is a polyhedron in three-dimensional space with vertices at* $(0,0,0)$, $(1,0,0)$, $(0,1,0)$ *and* $(1,1,n)$ *where n is a positive integer. Explain how the Reeve tetrahedron shows that any attempt to prove a simple version of Pick's theorem in three dimensions must fail.*

Answer □

Problem 74 (Bézout identity) *Two positive integers are said to be relatively prime if they have no factor in common. Given two relatively prime positive integers a and b, show that there exist positive or negative integers c and d such that*

$$ac + bd = 1.$$

Answer □

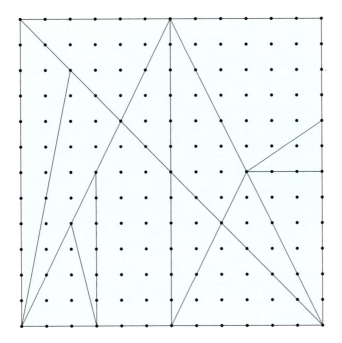

Figure 2.28: Archimedes's puzzle, called the Stomachion.

Problem 75 *Let T be a triangle with vertices at* $(0,0)$, $(1,0)$ *and* (m,n), *with m and n positive integers and* $n > 1$. *Must there be a grid point* (a,b) *in or on T other than one of the three vertices of T?* Answer □

2.7 Answers to problems

Problem 20, page 30

Figure 2.29 illustrates a number of points in the first quadrant that are (and are not visible) from the origin. Clearly $(1,1)$ is visible from the origin, but none of these points

$$(2,2),\ (3,3),\ (4,4),\ (5,5),\dots$$

(marked with an X in the figure) are visible precisely because $(1,1)$ *is in the way*. Similarly $(4,5)$ is visible from the origin but none of these points

$$(8,10),\ (12,15),\ (16,20),\ (20,25),\dots$$

are visible.

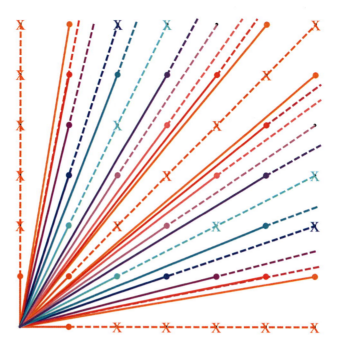

Figure 2.29: First quadrant unobstructed view from $(0,0)$.

The key observation here is the notion of *common factor*. You can prove (if you care to) that a point (m,n) on the grid is visible from the origin if and only if m and n have no common factors. (For example $(8,10)$ is not visible because both 8 and 10 are divisible by 2. Similarly $(12,15)$ is not visible because both 12 and 15 are divisible by 3. But $(4,5)$ is visible since no number larger than 1 divides both 4 and 5.)

In particular we see that some elementary number theory is entering into the picture quite naturally. That suggests that this investigation is perhaps not as frivolous and elementary as one might have thought. In Problem 74 we will see an application of Pick's theorem to number theory.

Problem 21, page 30

If you take the three points

$$(0,0),\ (1,1), (2,2)$$

as

$$V_1,\ V_2,\ V_3$$

then you will see the trouble we get into. We could avoid this with triangles by insisting that the three points chosen as vertices cannot lie on the same line.

Another example is taking

$$(0,0),\ (2,0),\ (1,1),\ (-1,1)$$

as
$$V_1, V_2, V_3, V_4.$$

Certainly there is a square with these vertices but we would have to specify a different order since the line segment V_1V_2 and the line segment V_3V_4 cross each other. We don't intend these to be the edges.

Yet again, an example taking
$$(0,0),\ (2,2),\ (2,0),\ (1,0,),\ (2,2),\ (0,0)$$

as
$$V_1, V_2, V_3, V_4, V_5, V_6$$

shows that we should have been more careful about specifying that the vertices are all different and the edges don't cross or overlap.

A reasonable first guess at a definition would have to include all the elements in the following statement:

2.7.1 *A polygon can be described by its vertices and edges that must obey these rules:*

1. *There are n distinct points*
$$V_1, V_2, V_3, \ldots, V_n.$$

2. *There are n straight line segments*
$$V_1V_2,\ V_2V_3, V_3V_4, \ldots, V_nV_1$$
   called edges. Two distinct edges intersect only if they have a common vertex, and they intersect only at that common vertex.

Even that is not quite enough for a proper mathematical definition, but will suffice for our studies. The reader might take this as a working definition that can be used in the solutions to the problems.

Problem 22, page 31

First consult your list to identify a vertex that occurs at a point (x,y) for which y is as large as possible. Then walk, without touching an edge, up to a vertex. Go around the polygon in order consulting your list of vertices for directions, staying close to the border, but without actually touching an edge or vertex. Eventually you will arrive near a vertex you have identified as having the largest y value. Which side of that point are you on?

This could be written up as a computer algorithm to test any point to find out whether it is inside or outside. Certainly in a finite number of steps (depending on how many edges we must follow) we can determine whether we are trapped inside or free to travel to much higher places.

Problem 23, page 32

There are five choices of splitting lines in addition to the line segment L. Notice that there are many other ways of joining pairs of vertices, but some ways produce line segments that are entirely outside the polygon or cross another edge. The six splitting choices are illustrated in Figure 2.7.

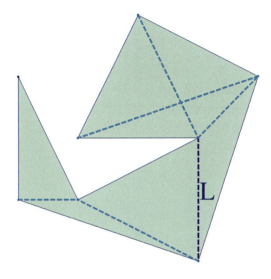

Figure 2.30: The six line segments that split the polygon.

Problem 24, page 32

Certainly you would have discovered quickly that no triangle can be split this way. But in every other case that you considered there would have been at least one line L that splits the polygon.

   Thus it appears to be the case that every polygon with four or more vertices can be split by some line segment that joins two vertices without passing through any other points on an edge of the polygon. That is the conjecture.

Problem 25, page 32

This may not be as obvious as it first appears, since we must consider all possible cases. It is easy to draw a few figures where many choices of possible vertices would not be allowed. It is clear in any particular example which two vertices can be used, but our argument must work for all cases.

   We assume that we have a polygon with n vertices where $n > 3$ and we try to determine why a line segment must exist that joins two vertices and is inside the polygon (without hitting another edge).

Go around the polygon's vertices in order until you find three consecutive vertices A, B and C such that the angle $\angle ABC$ in the interior of the polygon is less than 180 degrees. (Why would this be possible?)

The proof is now not too hard to sketch. Suppose first that the triangle ABC has no other vertices of the polygon inside or on it. If so simply join A and C.

The line segment AC cannot be an edge of the polygon. We know that AB and BC are edges. If AC were also an edge, then there are no further vertices other than the three vertices A, B, and C. Since we have assumed that there are more than 3 vertices this is not possible. (Statement 2.7.1 on page 65 has a formal description of a polygon that we can use to make this argument precise.) Consequently this line segment AC splits the polygon.

There may, however, be other vertices of the polygon in the triangle. Suppose that there is exactly one vertex X_1 in the triangle. Then, while AC cannot be used to split the polygon, the line segment BX_1 can. Again we are done. Suppose that there are exactly two vertices X_1 and X_2 in the triangle. Then one or both of the two line segments BX_1 or BX_2 can be used. To be safe choose the point closest to B.

Suppose that there are exactly three vertices X_1, X_2, and X_3 in the triangle. Then one or more of the three line segments BX_1 or BX_2 or BX_3 can be used. Draw some figures showing possible situations to see how this works. Note that the point closest to B is not necessarily the correct one to choose.

The general argument is a bit different. Suppose there are exactly n vertices X_1, X_2, ...X_n inside the triangle ABC. Select a point A' on the line AC that is sufficiently close to A so that the triangle $A'BC$ contains none of the points X_1, X_2, ...X_n. Now move along the line to the first point A'' where the triangle $A''BC$ does contain one at least of these points. From among these choose the vertex X_j that is closest to B. Then BX_j can be used to split the polygon since it can cross no edge of the polygon.

Problem 26, page 32

If M has m vertices, N has n vertices and the large polygon (before it was split) has p vertices then a simple count shows that

$$m+n = p+2$$

since the two endpoints of L got counted twice. But you can also observe that

$$m \geq 3 \text{ and } n \geq 3.$$

Combining these facts shows finally that

$$m = p+2-n \leq p+2-3 = p-1$$

and

$$n = p+2-m \leq p+2-3 = p-1.$$

So the two polygons *M* and *N* must have fewer vertices than the original polygon.

This fact will be a key to our induction proof later on. If every polygon (other than a triangle) can be split into subpolygons with fewer vertices, then we have a strategy for proving statements about polygons. Start with triangles (the case $n = 3$). Assume some property for polygons with 3, 4, ..., and n vertices. Use these facts to prove your statement about polygons with $n + 1$ vertices. Take advantage of the splitting property: the big polygon with $n + 1$ vertices splits into two smaller polygons with fewer vertices.

Problem 27, page 33

Perhaps you answered that this was the only triangulation possible. If so you didn't look closely enough. There is one more triangulation of *P* that uses additional edges joining a pair of vertices as Figure 2.7 illustrates.

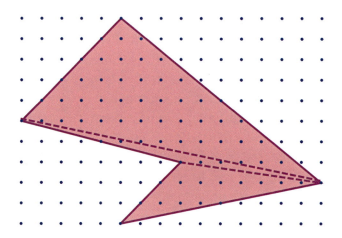

Figure 2.31: Another triangulation of *P*.

But, in fact, any decomposition of *P* into smaller triangles would also be considered a triangulation and can be used to compute areas. The most interesting triangulations for our study of polygons on a grid might require us to use grid points for vertices of the triangles. There are many such triangulations possible for *P*.

More generally still, we could ignore the grid points entirely and allow any decomposition into smaller triangles. Once again, there are many such triangulations possible for *P*; indeed there are infinitely many.

Problem 28, page 33

To start the problem try finding out why a polygon (of any shape) with four vertices can always be triangulated. Then work on the polygon with five vertices but use the splitting argument to ensure that this polygon can be split into smaller polygons, each of which is easy to handle.

A complete inductive proof for the general case is then fairly straightforward. Let n be the number of vertices of a polygon P. If $n = 3$ then the polygon is already triangulated. If $n = 4$ simply join an appropriate pair of opposite vertices and it will be triangulated. If $n = 5$ use the splitting argument (which we have now proved in Problem 25) to split P into smaller polygons. Those small polygons have 3 or 4 vertices and we already know how to triangulate them. And so on . . . .

Well "and so on" is not proper mathematical style. But this argument is easy to convert into a proper one by using the mathematical induction. You may need to review the material in the appendix before writing this up.

Problem 30, page 36

It is good practice in starting a topic in mathematics to experiment on your own with the ideas and try out some examples. All too often in a mathematics course the student is copying down extensive notes about definitions and theorems well before he is able to conceptualize what is happening.

In this case you will certainly have computed polygons with some or all of these areas:

$$\frac{1}{2}, \ 1, \ \frac{3}{2}, \ 2, \frac{5}{2}, \ 3, \ \frac{7}{2}, \ 4, \ldots.$$

But you will not have found any other area values. We certainly expected fractions, but why such simple fractions? All areas appear to be given by some formula

$$\frac{N}{2}$$

where N is an integer. This, if it is true, is certainly a remarkable feature of such figures. Few of us would have had any expectation that this was going to happen.

Our best guess is that, for polygons on square grids, something is being counted and each thing counted has been assigned a value that is a multiple of $1/2$. The natural thing we might consider counting is grid points. But what values should we assign to each grid point?

Problem 31, page 36

As we have already determined, a triangle T with vertices at $(0,0)$, (s,t), and (a,b) must have area given by

$$\left| \frac{at - bs}{2} \right|.$$

The numerator is an integer so the area is clearly a multiple of $1/2$.

Now, by drawing some pictures, try to find an argument that allows you to conclude that all triangles anywhere on the grid can be compared to a triangle like this. We must be able to claim that every triangle on the grid is congruent to one with a vertex at $(0,0)$ and of this type. Then, since we have determined that this triangle has area an integer multiple of $1/2$, then every triangle on the grid has this property.

Problem 32, page 36

In Problem 28 we saw that all polygons on the grid can be triangulated by triangles on the grid. Each such triangle has an area that is a multiple of $1/2$. The polygon itself, being a sum of such numbers, also has an area that is a multiple of $1/2$.

Problem 33, page 42

You should be able to compute easily the area of any triangle that has one side that is horizontal or one side that is vertical. In that case the formula

$$1/2 \times \text{base} \times \text{height}$$

immediately supplies the answer. For primitive triangles of this type you will observe that both base and height are 1 so the area is immediately $1/2$.

If the triangle has no side that is horizontal or vertical then the formula

$$1/2 \times \text{base} \times \text{height},$$

while still valid, does not offer the easiest way to calculate the area. For these triangles the methods of Section 2.1.6 should be used. For example compute the area of the primitive triangle with vertices at $(0,0)$, $(2,1)$ and $(3,2)$. Try a few others.

You should have found that all of them that you considered have area exactly $1/2$. Again the number $1/2$ emerges and seems (perhaps) to be related to the fact that all of these figures have exactly three points on the grid. Also, we know that every triangle on the grid has an area that is some multiple of $1/2$; since primitive triangles are somehow "small" we shouldn't be surprised if all have area exactly $1/2$, the smallest area possible for a triangle on the grid.

Problem 34, page 42

You should have found that all of them have area exactly 1. We can compare with primitive triangles in a couple of ways. Problem 33 shows that primitive triangles must have area $1/2$.

The extra grid point on the edge of these triangles appears to contribute an extra credit of $1/2$. Or, perhaps, we could observe that the extra grid point allows us to split the triangle into two primitive triangles each of which has area 1/2. Both viewpoints are useful to us.

Problem 35, page 42

You should have found that all of them have area exactly 3/2. The single grid point in the interior of the triangle can be joined to the three vertices, dividing the original triangle into three primitive triangles. Since each of these has area 1/2 according to Problem 33, the total area is 3/2.

Problem 36, page 42

You should have found that all of them have area exactly 7. It is likely a mystery to you, however, whether these two numbers 4 and 6 adequately explain an area of 7. (Is there some formula for which, if you input 4 and 6, the result will be 7?)

Does this mean that all such polygons (with 4 boundary points and 6 interior points) must have area 7? Our choice of polygons was driven mostly by a desire to find figures whose area could be computed without much difficulty. It is not clear at this stage whether much weirder figures would or would not have this property.

But, if this is so, then it appears (quite surprisingly) to be the case that the area inside a polygon with vertices on the grid depends only on knowing how many grid points there are on the polygon itself and how many grid points there are inside the polygon.

Problem 37, page 42

In Problem 30 you likely constructed a few of these. Just describe a procedure that would construct one example for each of these. Start perhaps with a triangle with vertices at $(0,0)$, $(0,1)$, and $(1,0)$. Just keep adding simple primitive triangles until you see a way to write up your recipe.

Problem 38, page 42

Your experiments should have produced squares with these areas:

$$1, 2, 4, 5, 9, 10, 13, 16, 17, 20, 25, 26, 29, 36, 37, 40, 45, 52\ldots.$$

If you didn't find many of these keep looking before you try to spot the pattern or try to explain the pattern.

Certainly, for any integer k, the squares with vertices $(0,0)$, $(0,k)$, $(k,0)$ and (k,k) is on the grid and has area k^2. This explains all of these numbers:

$$1, 4, 9, 16, 25, 36, 49, 64, 81, 100,\ldots.$$

What about the other numbers in the list we found above?

But the square with vertices $(0,0)$, $(1,1)$, $(-1,1)$ and $(2,0)$ also works and has area 2 since each side length is $\sqrt{2}$. That explains the number 2. More generally, for any choice of point (a,b), there is a square with one vertex at $(0,0)$ and the line joining $(0,0)$ to (a,b) as one of its sides. The side length is

$$\sqrt{a^2+b^2}$$

by the Pythagorean theorem and so the area is

$$a^2+b^2.$$

Consequently any number that is itself a square or is a sum of two squares must be the area of a square on the grid. That statement describes the list of possibilities that we saw.

Problem 39, page 42

The area of the rectangle R is 8. The number of grid points inside the rectangle is 3. Thus counting grid points inside is a considerable underestimate in this case. Perhaps, however, with much larger rectangles this might be a useful first estimate.

Similarly, the area of the triangle T is 4. The number of grid points inside T is 1. Again simply counting grid points inside gives too low an estimate.

You may wish to try some other examples and see if the same kind of conclusion is reached. A simple counting of grid points inside produces estimates that are poor for these relatively small polygons.

Problem 40, page 43

Once again the area of the rectangle R is 8. The number of grid points inside P is 3 to which we are instructed to add the number of grid points on the rectangle itself. There are 12 such points and adding these gives $12+3=15$, considerably larger than 8.

The area of the triangle T is 4. The number of grid points inside T is 1 and the number of grid points on the triangle is 8. The addition is $1+8=9$, rather more than the area of the triangle.

It appears that, in order to reduce the total Pick credit so that it is closer to the actual areas we need to give less credit to some of the points.

Problem 41, page 43

The grid points on R and T are neither inside the polygon nor outside. We can try giving them less credit than 1. Our choices are 0 and $1/2$.

Let's try $1/2$ for all of them which would be a reasonable first guess. For R we find 12 such points (counting the corners of R). Giving each such point half

credit we obtain
$$3+6=9$$
whereas the area of R is 8. This is rather closer but is just an overestimate by 1.

Similarly, for the triangle T there are 8 grid points on the triangle. If we give them half-credit, we obtain
$$1+4=5.$$
The area of the triangle is 4 and so, once again, we have found an overestimate by exactly 1.

Try some other figures to see if this is what will always happen. Should we change the credit (reduce some of these points to zero credit) or should we try to figure out why the extra 1 arises?

Problem 42, page 43

Your examples should show results similar to those we found in Problem 41. Trying for an estimate
$$\text{Area}(P) \approx [\text{\# of grid points inside } P] + \frac{[\text{\# of grid points on } P]}{2}$$
using our idea of full credit 1 for the inside points and half-credit $1/2$ for the boundary points, in each case we found an overestimate by one unit. Did you?

Problem 43, page 45

We have already seen that the formula
$$\text{Area}(P) = [\text{\# of grid points inside } P] + \frac{[\text{\# of grid points on } P]}{2} - 1$$
works in a few simple cases. Let us check that it must always work for rectangles with vertical and horizontal sides and with vertices on the grid work.

If the rectangle R has dimensions m and n the actual area is the product mn. We can count directly that
$$[\text{\# of grid points inside } R] = (m-1)(n-1).$$
and
$$[\text{\# of grid points on } R] = 2(m+n).$$
(Check these.)

Thus our calculation using this formula would result in
$$(m-n)(n-1) + \frac{2(m+n)}{2} - 1 = mn.$$
Since this is the correct area of the rectangle, the formula is valid at least in this special case.

Problem 44, page 45

We have already seen that the formula

$$\text{Area}(P) = [\text{\# of grid points inside } P] + \frac{[\text{\# of grid points on } P]}{2} - 1$$

works for all rectangles and, in a few simple cases, for some triangles. Let us show that it works if $P = T$ is a triangle with one vertical side and one horizontal side and with vertices on the grid work.

If the horizontal and vertical sides have length m and n, the area of the triangle is $mn/2$. Adjoining another triangle T' as we did in Figure 2.14 we arrive at a rectangle R whose area is mn that is split into the two triangles T and T'. The two triangles are identical (one is a reflection of the other) and so they have the same areas and the same number of grid points inside and on the boundary.

Let p be the number of grid points on the diagonal of the rectangle, excluding the two vertices. (There may be none.) We easily compute (using Figure 2.14 as a guide)

$$[\text{\# of grid points inside } T] + [\text{\# of grid points inside } T'] + p$$

$$= [\text{\# of grid points inside } R]$$

and

$$[\text{\# of grid points on } T] + [\text{\# of grid points on } T']$$

$$= [\text{\# of grid points on } R] + 2 + 2p.$$

This last identity is because the two vertices on the diagonal are counted twice, once for T and once for T' as also are any of the other p grid points on the diagonal. Thus we can check using simple algebra that

$$2 \times \left\{ [\text{\# of grid points inside } T] + \frac{[\text{\# of grid points on } T]}{2} - 1 \right\}$$

$$= [\text{\# of grid points inside } R] + \frac{[\text{\# of grid points on } R]}{2} - 1 = mn.$$

This last identity is clear since we already know that our formula works to compute the area of any rectangle, and here R has area mn.

Thus we have verified that the formula does produce exactly $mn/2$, which is the correct area for the triangle T. This handles triangles, but only (so far) those oriented in a simple way with a horizontal side and a vertical side.

The algebra is not difficult but it does not transparently show what is going on. In Section 2.3.3 we explore this in a way that will help considerably in seeing the argument and in generalizing it to more complicated regions.

Problem 45, page 45

The count is quite easy to do. Except for points on the line L every point in the count for Pick(P) is handled correctly in the sum. The points on L, however, all get counted twice. The two vertices at the ends of L get a count of $1/2 + 1/2$ in the count for Pick(P) but they get $1/2 + 1/2 + 1/2 + 1/2$ for the count Pick(M) + Pick(N). So that is 1 too much.

What about the remaining grid points, if any, on L? They are also counted twice. But this takes care of itself. In the count for Pick(M) + Pick(N) any such point gets a count of $1/2 + 1/2$. But that is exactly what it receives in the count for Pick(P) since it is now an interior point and receives credit of 1. In short then, without much trouble, we see that

$$\text{Pick}(M) + \text{Pick}(N) = \text{Pick}(P) + 1$$

where the extra 1 is explained simply by the fact the endpoints of the edge L got counted twice.

Problem 46, page 45

We want to prove that

$$\text{Area}(T) = \text{Pick}(T) - 1$$

for any triangle with horizontal and vertical sides. As we did in our previous solution we introduce T' the mirror image of T so that T and T' together form a rectangle R. Then

$$\text{Pick}(T) = \text{Pick}(T'),$$

$$\text{Pick}(T) + \text{Pick}(T') = \text{Pick}(R) + 1,$$

and

$$2\,\text{Area}(T) = \text{Area}(R)$$

We are allowed to use the fact that $\text{Area}(R) = \text{Pick}(R) - 1$ that we proved earlier. So

$$\text{Area}(T) = \frac{\text{Area}(R)}{2} = \frac{\text{Pick}(R) - 1}{2} = \frac{2\,\text{Pick}(T) - 1 - 1}{2} = \text{Pick}(T) - 1$$

which is the formula we wanted.

Problem 48, page 47

This is just a warm-up to the general case discussed in Problem 49. It is worth trying to handle this one using the ideas developed so far since some thinking on this problem helps understand better what is needed for the harder problem.

For example, if the triangle is obtuse angled like the triangle T in Figure 2.32 then add a right-angled triangle P so that T and P together make another right-

angled triangle Q. We know already that
$$\text{Area}(P) = \text{Pick}(P) - 1$$
and
$$\text{Area}(Q) = \text{Pick}(Q) - 1$$
but we want to show that
$$\text{Area}(T) = \text{Pick}(T) - 1$$
is also valid. Simply use $\text{Pick}(T) + \text{Pick}(P) = \text{Pick}(Q) + 1$ and $\text{Area}(T) + \text{Area}(P) = \text{Area}(Q)$.

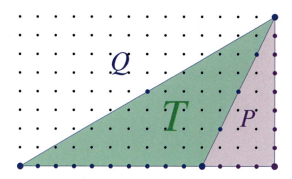

Figure 2.32: Obtuse-angled triangle T with a horizontal base.

If the triangle is acute-angled like the triangle T in Figure 2.33 then it can be split into two right-angled triangles and handled in a similar way.

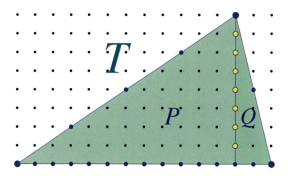

Figure 2.33: Acute-angled triangle T with a horizontal base.

Problem 49, page 47

Let R be the smallest rectangle with horizontal and vertical sides that contains T. Then R is comprised of T and some other polygons for which we have already established the Pick formula. Figure 2.34 illustrates how the triangle T plus some other simpler triangles, and possibly a rectangle, might make up the whole of the rectangle.

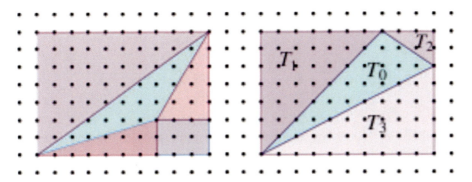

Figure 2.34: Triangles whose base is neither horizontal nor vertical.

Note the similarity between Figure 2.34 and Figure 2.6. Apply reasoning similar to that used in Problem 29 to determine whether the formula is valid for an arbitrary triangle. This suggestion should enable you to solve the problem. There is a detailed discussion, in any case, in Section 2.4.1.

Problem 50, page 48

The algebra is quite simple, just a lot of adding and subtracting. Here is what we know:

$$\text{Area}(R) = \text{Area}(T_0) + \text{Area}(T_1) + \text{Area}(T_2) + \text{Area}(T_3),$$
$$\text{Pick}(R) + 3 = \text{Pick}(T_0) + \text{Pick}(T_1) + \text{Pick}(T_2) + \text{Pick}(T_3),$$
$$\text{Area}(R) = \text{Pick}(R) - 1,$$
$$\text{Area}(T_1) = \text{Pick}(T_1) - 1,$$
$$\text{Area}(T_2) = \text{Pick}(T_2) - 1,$$

and

$$\text{Area}(T_3) = \text{Pick}(T_3) - 1.$$

Thus

$$\text{Area}(T_0) = \text{Area}(R) - \{\text{Area}(T_1) + \text{Area}(T_2) + \text{Area}(T_3)\}$$
$$= \text{Pick}(R) - 1 - \{\text{Pick}(T_1) + \text{Pick}(T_2) + \text{Pick}(T_3) - 3\}$$
$$= \{\text{Pick}(R) - \text{Pick}(T_1) - \text{Pick}(T_2) - \text{Pick}(T_3)\} + 2.$$

But

$$\text{Pick}(R) + 3 = \text{Pick}(T_0) + \text{Pick}(T_1) + \text{Pick}(T_2) + \text{Pick}(T_3),$$

which is the same as

$$\text{Pick}(R) - \text{Pick}(T_1) - \text{Pick}(T_2) - \text{Pick}(T_3) = \text{Pick}(T_0) - 3.$$

Finally then

$$\text{Area}(T_0) = \{\text{Pick}(T_0) - 3\} + 2 = \text{Pick}(T_0) - 1.$$

The proof is complete.

Problem 51, page 48

Figure 2.35 (which is just a repeat of Figure 2.1 in the text) indicates rather well which grid points to use. As you can see, there are six points on the boundary (in addition to the five vertices) that must be included in our accounting. For the second half of the problem, triangulate into just three convenient triangles and check the areas of each by counting according to the Pick formula that we have now verified for triangles.

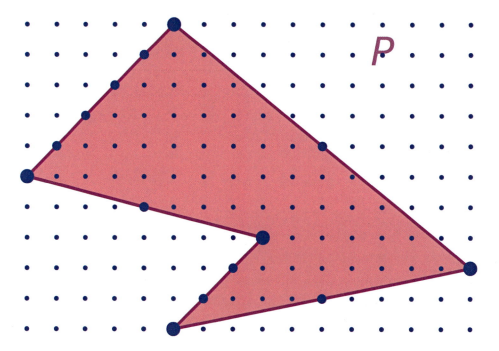

Figure 2.35: What is the area inside *P*?

Problem 52, page 49

Let us set up an argument using mathematical induction. For each integer $k \geq 3$ let $P(k)$ be the statement that for every polygon with k or fewer sides the formula works. We already know $P(3)$ is valid (the formula is valid for all triangles)

Now suppose the formula is valid for all polygons with n or fewer sides. (This is the induction hypothesis.) Let P be any polygon with $n+1$ sides. We must show the formula is valid for P.

At this point we need the splitting argument. The essential ingredient in all inductive proofs is to discover some way to use the information in the induction hypothesis (in this case the area formula for smaller polygons) to prove the next step in the induction proof (the area formula for the larger polygon).

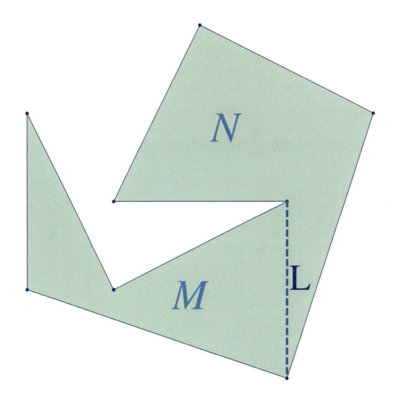

Figure 2.36: Finding the line segment L.

As in Figure 2.36 we use the splitting argument to find a line segment L whose vertices are endpoints of P and the rest of L is inside P. In the figure the line segment L has separated P into two polygons M and N. Because we have added L, the total number of sides for M and N combined is now $n+2$, but each of the polygons separately has fewer than $n+1$ sides. Thus, by the induction hypothesis, the formula is valid for each of the polygons M and N.

Thus we know that

$$\text{Area}(P) = \text{Area}(M) + \text{Area}(N),$$

while

$$\text{Area}(M) = \text{Pick}(M) - 1$$

and

$$\text{Area}(N) = \text{Pick}(N) - 1.$$

By our additivity formula for the Pick count,

$$\text{Pick}(P) + 1 = \text{Pick}(M) + \text{Pick}(N).$$

Simply putting these together gives us

$$\text{Area}(P) = \text{Area}(M) + \text{Area}(N) = \text{Pick}(M) - 1 + \text{Pick}(N) - 1$$
$$= \text{Pick}(P) + 1 - 2 = \text{Pick}(P) - 1.$$

This verifies that the Pick formula works for our polygon P with $n+1$ sides. This completes all the induction steps and so the formula must be true for polygons of any number of sides.

Problem 53, page 53

Figure 2.37 shows a possible ending position for this game. There are no further moves possible.

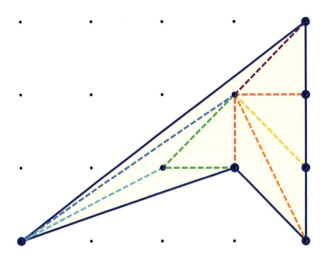

Figure 2.37: A final position in this game.

One thing that is evident from this particular play of the game is that the final position is a primitive triangulation of the original polygon. Would all plays of the game result in a primitive triangulation?

In playing this game there were exactly 8 moves and so it was the second player who made the last move and won the game. Would all plays of this game have the same result or was the second player particularly skillful (or lucky)?

Problem 54, page 53

Choose a polygon that is not too large and play a few games (alone or with a friend). You will certainly observe that the game ends with a primitive triangulation of the original polygon. You may also have observed that, if you lost the game, each time you repeated the game (with the same starting position) you also lost, no matter what new strategy you tried.

Did you observe anything else? You could have, if you thought of it, also have counted the number of moves and counted the number of triangles in the final figure. But perhaps you didn't notice anything about this count beyond the fact that the number of moves and the number of triangles are closely related and these numbers didn't change when you replayed the game on this polygon.

Problem 55, page 53

The game ends after a certain number of moves. Call this number n. Thus, after n moves, the polygon has been split into $n+1$ subpolygons.

Are they all triangles?

Let us suppose not, i.e., that there is a subpolygon in the final position with 4 or more vertices. According to the splitting argument of Section 2.1.4 there must be a line segment joining two of these vertices for which the line segment is entirely inside the subpolygon. But that would allow a Type 1 move to be made and so the game is not over after all.

Since they are all triangles we can ask

"Are they all primitive triangles?"

Suppose T is a triangle in the final position. Does T have a grid point on the boundary other than the three vertices? If it did, then clearly a Type 1 move could have been made by joining that point to an opposite vertex. Does T have an interior grid point? If it did, then clearly a Type 2 move could have been made by joining that point to two of the vertices. This shows that each triangle in the final triangulation must be primitive.

Problem 56, page 53

Recall that a triangle with vertices on the grid is said to be primitive if the only grid points on or in the triangle are the three vertices themselves. What is the area of a primitive triangle?

Not surprisingly the answer is $1/2$. We know that all polygons on the grid have an area that is a multiple of $1/2$. These are the smallest such polygons. We have also experimented in a few instances with primitive triangles (e.g., in Problem 33) and in each case we found an area of $1/2$.

The Pick formula supplies this immediately. If T is a primitive triangle, then there are no interior grid points ($I = 0$) and there are only the three boundary grid points ($B = 3$). Consequently

$$\text{Area}(T) = I + B/2 - 1 = 0 + 3/2 - 1 = 1/2.$$

as we would have suspected.

Problem 57, page 53

Consider the final position. After n moves the polygon has been split into $n+1$ subpolygons, each of which we now know (because of Problem 55) is a primitive triangle.

Each primitive triangle has area $1/2$ (by Pick's rule) and so the area of the original polygon P must be

$$\text{Area}(P) = \frac{n+1}{2}$$

since there are $n+1$ primitive triangles. Pick's theorem says, on the other hand, that

$$\text{Area}(P) = I + B/2 - 1.$$

Comparing these two expressions we see that

$$\frac{n+1}{2} = \frac{2I+B-2}{2}$$

which shows that the number of primitive triangles in the final configuration is

$$n+1 = 2I + B - 2.$$

This number is always the same even though there may be a great many different ways of ending up with a primitive triangulation.

The number of moves in the game is then always given by

$$n = 2I + B - 3.$$

Problem 58, page 53

In Problem 57 we determined that, no matter what strategy either player elects to try, the number of moves in the game is always given by

$$n = 2I + B - 3.$$

The first player wins if this is odd. The second player wins if this is even.

Looking again at that number it is evident that the first player wins simply if B is even and the second player wins if B is odd. The number of interior points I is irrelevant to the question of who wins (although the game is much longer if I is big).

So the game is rigged. The player in the know just offers to go second in a game if she spots that B is odd.

Problem 59, page 55

Let us play the game on the polygon P splitting it by a Type 1 move into two polygons M and N. For a Type 1 move there is a line L joining two grid points on the boundary of P that becomes a new edge for M and N. We consider both sides of the equation

$$\text{Count}(P) = \text{Count}(M) + \text{Count}(N)$$

that we wish to prove. Draw a picture or else what follows is just words that may not convey what is happening.

Now

$$\text{Count}(P) = 2I + B - 2,$$
$$\text{Count}(M) = 2I_M + B_M - 2,$$

and

$$\text{Count}(N) = 2I_N + B_n - 2$$

We use a simple counting argument looking at the grid along the splitting line L. The count works out perfectly for points that are not on the splitting line. Every point in the B count appears in the counts for B_M or B_N; every point in the I count appears in the counts for I_M or I_N.

For the grid points that are on the splitting line L, the two endpoints of L are counted twice, once for for B_M and once for B_N. The extra -2 accounts for that. Any interior grid points on L that appeared in the count for I (where they count double) now appear in the counts for B_M and B_N (where they count as 1). That takes care of them too.

There remains only to do the same for a Type 2 move. But really the same argument applies without any changes.

Problem 60, page 55

There are a number of ways to do this. One cute way is to use the primitive triangulation result itself to do this. The idea is that we already know primitive triangles have area at least $1/2$. (See Problem 31.) A clever triangulation will show that they cannot possibly have area more than $1/2$.

Take any rectangle R on the grid with horizontal and vertical sides. We suppose the rectangle has dimensions $p \times q$. Thus

$$\text{Area}(R) = pq.$$

We can easily count interior points and boundary points for such a rectangle.

We triangulate the rectangle so as to find a primitive triangulation of R. But we know how many primitive triangles there must be for R: we just need to compute that

$$B = 2p + 2q$$

while

$$I = (p-1)(q-1).$$

So if n is the number of primitive triangles our formula gives us

$$n = \text{Count}(R) = 2I + B - 2 = 2(p-1)(q-1) + 2p + 2q - 2 = 2pq.$$

All of our triangles have area at least $1/2$ so if any one of them has area more than $1/2$ the area of the rectangle would be bigger than pq which is impossible. Thus each has area $1/2$.

Every primitive triangle can appear somewhere inside such a rectangle and be used in a primitive triangulation, so this argument applies to any and all primitive triangles.

Problem 61, page 55

Well many mathematicians would. But there is a huge intuitive leap from a problem about area to a problem about primitive triangulations. We began early on to sense a connection and finally came to a full realization only later on.

  We could have simply announced the connection and then pursued this line of argument. Plenty of mathematics textbooks and lectures do this kind of thing all the time. The proofs are fast, slick, and the student's intuition is left behind to catch up later. For a book on Mathematical Discovery we can take our time and try to convey some idea of how new mathematics might be discovered in the first place.

Problem 62, page 56

In Figure 2.23 we can measure the rectangles directly and see that P is a 5×12 rectangle and H is a 2×4 rectangle. Thus the area of the region G between P and H must be

$$\text{Area}(G) = \text{Area}(P) - \text{Area}(H) = 60 - 8 = 52.$$

[We could have used, instead, our old method of counting interior points at full value of 1 and points on the polygon at half value of $1/2$. For P we count 44 interior points and 34 points on P. Thus our standard formula gives

$$\text{Area}(P) = 44 + 34/2 - 1 = 44 + 17 - 1 = 60$$

as expected. For H we have 3 interior points and 12 points on H so

$$\text{Area}(H) = 3 + 12/2 - 1 = 3 + 6 - 1 = 8$$

which is again correct.]

  Let's see what we get if we try to use our formula for the area of the region G between P and H. Here, once again, G has interior points and points on the boundary; all the points that are on the boundary of H must be considered on the boundary of G.

  We note that the grid points of the interior of G consists of those inside P except the 15 that lie inside or on H. There are 29 such points so $I = 29$. The boundary of the region in question consists of the polygons P and H. There are

$$B = 34 + 12 = 46$$

grid points on this boundary. Trying our usual computation for G, we obtain

$$\text{Area}(G) = I + B/2 - 1 =$$
$$29 + 46/2 - 1 = 51?$$

This is actually quite encouraging since our formula gave us a result that is *too small* by only one unit.

Try some other choices for P and H. Both should be polygons with vertices at grid points, H should be inside P, and G is the region formed from subtracting H and its inside from the inside of P. Rectangles (as we used) make for the simplest computations. Try triangles and a few others.

Problem 63, page 56

Let's argue as we have several times previously. The grid points inside P are of three types: those that are inside H, those that are on H, and those that are not inside H nor on H. Our computation for the area inside G gave zero credit for the first type of point, half credit to the second type of point, and full credit to the third type of point. It also gave half credit for the grid points on P.

Thus the total credit given to G is provided by the area formula

$$\text{Area}(G) = I + B/2$$

whereas Pick's formula (for polygons *without* holes) would be

$$I + B/2 - 1$$

instead, resulting in too low a number for the area.

Problem 64, page 57

Simple algebra gives

$$\text{Area}(P) - \text{Area}(H) = I(P) - I(H) + [B(P) - B(H)]/2. \qquad (2.1)$$

Now figure out what $I(G)$ and $B(G)$ must be. Directly we can see that $I(G)$ includes the points counted for $I(P)$ excluding those counted in $I(H)$ as well as those counted in $B(H)$. Thus

$$I(G) = I(P) - I(H) - B(H).$$

Similarly we can see that $B(G)$ includes all of the points counted for $B(P)$ plus those counted in $B(H)$. Thus

$$B(G) = B(P) + B(H).$$

Put this altogether using elementary algebra and find that

$$\text{Area}(G) = \text{Area}(P) - \text{Area}(H)$$
$$= I(P) - I(H) + [B(P) - B(H)]/2$$
$$= [I(P) - I(H) - B(H)] + [B(P) + B(H)]/2 = I(G) + B(G)/2.$$

So finally the new formula for the region G (i.e., P with a hole H) is

$$\text{Area}(G) = I(G) + B(G)/2$$

which is exactly Pick's formula without the extra -1. This is what we have already observed for specific examples except that now we have an algebraic proof of this fact.

We can think of this using the phrase "without the extra -1" or we could write our new one-hole formula as

$$\text{Area}(G) = \left(I(G) + \frac{B(G)}{2} - 1\right) + 1$$

which might be more helpful since it asks us to add 1 to Pick's formula.

Problem 65, page 57

The formula for the area G that remains inside a polygon with exactly n polygonal holes is

$$\text{Area}(G) = \left(I(G) + \frac{B(G)}{2} - 1\right) + n.$$

Note that $n = 0$ (i.e., no holes) is exactly the case for Pick's Rule and so our new formula is a generalization of Pick's original formula.

The proof can be argued via counting, as we have done often, or algebraically as in our last proof. Here $B(G)$ is the count we obtain for all points lying on P as well as on any of the n polygons that create the holes. (We assume no two of the polygons have points in common).

We leave the details to the reader, but for those who are interested, we provide a calculation for the case of two holes. Suppose P is a polygon with holes created by two smaller polygons Q and R as in Figure 2.38.

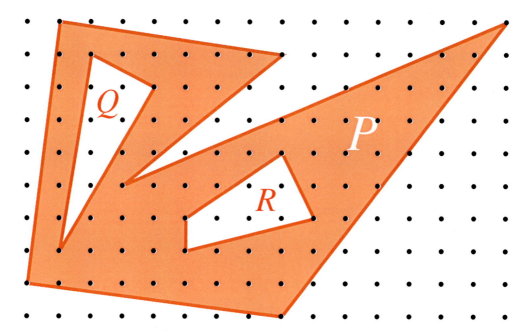

Figure 2.38: Polygon with two holes.

We show that $I(G) + B(G)/2$ is one less than $A(G)$. We have

$$I(G) = I(P) - I(Q) + B(Q) - I(R) - B(R)$$

and
$$B(G) = B(P) + B(Q) + B(R).$$

Thus

$$I(G) + B(G)/2 =$$
$$I(P) - I(Q) - I(R) - B(Q) - B(R) + [B(P) + B(Q) + B(R)]/2 =$$
$$I(P) - I(Q) - I(R) + B(P)/2 - [B(Q) + B(R)]/2 =$$
$$I(P) + B(P)/2 - [I(Q - B(Q)]/2 - [I(R) + B(R)]/2 =$$
$$\text{Area}(P) + 1 - [\text{Area}(Q) + 1] - [\text{Area}(R) + 1] = A(G) - 1.$$

Thus the correct formula for the area of G is

$$\text{Area}(G) = [I(G) + B(G)/2 - 1] + 2$$

as was to be shown.

For those of our readers rather braver here is the proof for the general case. It is exactly the same but just needs some extra attention to notation so that the task of adding up n different elements is not so messy.

Instead of labeling the smaller polygons as Q, R, ... let us call them P_1, P_2, ..., P_n and let us call the big polygon P_0. Write $A_i = A(P_i)$, $B_i = B(P_i)$, and $I_i = I(P_i)$. Then, for each $i = 0, 1, 2, \ldots, n$ we know that Pick's Rule provides

$$A(P_i) = I_i + B_i/2 - 1$$

and so, if G is the figure P_0 with all the holes removed, then

$$A(G) = A(P_0) - \sum_{i=1}^{n} A(P_i) =$$

$$I_0 + \frac{B_0}{2} - 1 - \sum_{i=1}^{n} \left[I_i + \frac{B_i}{2} - 1 \right].$$

But it is easy to check that

$$I(G) = I_0 - \sum_{i=1}^{n} (I_i + B_i)$$

and

$$B(G) = B_0 + \sum_{i=1}^{n} B_i.$$

Put these together to obtain the final formula

$$A(G) = [I(G) + B(G)/2 - 1] + n$$

as was to be shown.

Problem 66, page 60

This is almost obvious. For the points on the common edge, the angle of visibility for P_1 and the angle of visibility for P_2 add together to give the angle of visibility for P. For every other point there is no problem since they can appear only in the count for P_1 or else in the count for P_2.

Problem 67, page 60

Start with triangles exactly as before and show that

$$\text{Area}(T) = \text{Pick}^*(T)$$

for every triangle. This takes a few steps as we have already seen in Section 2.4.1.

Then, since the new Pick count is strictly additive (no extra 1 to be added), any figure that can be split into triangles allows the same formula for the area. But any polygon can be triangulated.

Problem 68, page 60

For each point in or on a polygon P with a number of holes H_1, H_2, ...H_k we decide what is its *angle of visibility*. This is the perspective from which standing at a point we see into the inside of the region. For points interior to the region we see a full 360 degrees. For points on an outer edge of P but not at a vertex we see only one side of the edge, so the angle of visibility is 180 degrees. The same is true for points on an edge of a hole, but not a vertex of the hole.

For points at an outer vertex of the region the angle of visibility would be the interior angle and it could be anything between 0 degrees and 360 degrees. Finally for points on the boundary of the region that are vertex points of one of the holes we do the same thing. One side of the angle looks into the hole, the other side looks into the region of concern.

As before our *modified Pick's count* is to take each possible grid point into consideration, compute its angle of visibility, divide by 360 to get the contribution. Points inside get 360/360=1. Points on the edge but not at a vertex get 180/360=1/2. And, finally, points at the vertex get $a/360$ where a is the degree measure of the angle. The new Pick count we will write as

$$\text{Pick}^*(P).$$

Now, using G to denote the region defined by removing the holes from inside of P, simply verify that

$$\text{Pick}^*(G) + \text{Pick}^*(H_1) + \text{Pick}^*(H_2) \cdots + \text{Pick}^*(H_k) = \text{Pick}^*(P).$$

This is far easier than it appears. The only points that get counted twice in the sum on the left side of the equation are points on the boundary of G that are also on a particular hole H_i. In computing $\text{Pick}^*(H_i)$ that point gets a count of $a/360$ where the a is the angle interior to H_i. In computing $\text{Pick}^*(G)$ that same point gets a count of $[360-a]/360$. The sum is 1 which the correct value for this point since, considered in P itself it is an interior point.

The rest of the proof is obvious and requires only that we use

$$\text{Area}(G) = \text{Area}(P) - \text{Area}(H_1) - \text{Area}(H_2) - \cdots - \text{Area}(H_k) = \text{Pick}^*(G).$$

This uses the fact that we know this formula for all polygons without holes.

Problem 69, page 60

Problem 68 presented an easier and more intuitive proof of a formula for the area of a polygonal region G with n holes. We need to relate it to the other formula.

Use G to denote the region defined by removing a number of holes H_1, H_2, $\ldots H_n$ from inside of P, and use $\text{Pick}^*(G)$ to represent the count that uses the angle of visibility.

Use $\text{Pick}(G)$ to represent the simpler count

$$\text{Pick}(G) = I + B/2$$

wherein the number of boundary points B of G must include points on the boundary of P as well as on the boundary of one of the holes. The number I, as usual, counts the number of interior points (here these are points inside P but not in one of the holes).

Now simply show that

$$\text{Pick}^*(G) = \text{Pick}(G) - 1 + n.$$

That explains Pick's formula and illustrates where the n appears.

To verify this equation we need only focus on the vertices of one of the holes H_i. Every other point is counted the same whether it appears in the count for $\text{Pick}^*(G)$ or the count for $\text{Pick}(G)$.

If there are h vertices on that hole H_i then we recall that the interior angles (interior to the hole H_i) would have a sum

$$a_1 + a_2 + \cdots + a_h = 180(h - 2).$$

since the angles inside any polygon with h vertices add up to $180(h-2)$ degrees.

But in the computation for $\text{Pick}^*(G)$ the same angles at the vertices appear but are complementary, i.e., the corresponding angles are

$$(360 - a_1), \ (360 - a_2), \ldots, (360 - a_h).$$

Thus we can compute the contributions of the vertices of the hole H_i to the count for $\text{Pick}^*(G)$ to be

$$\frac{(360 - a_1) + (360 - a_2) + \cdots + (360 - a_h)}{360}$$

$$= \frac{360h - [a_1 + a_2 + \cdots + a_h]}{360}$$

$$= \frac{360h - 180(h - 2)}{360}$$

$$= h/2 + 1.$$

The count for the computation of Pick(G) using these same vertices is simply $h/2$, which is one smaller. But that is one smaller *for each hole*. This verifies

$$\text{Pick}^*(G) = \text{Pick}(G) - 1 + n$$

and explains the appearance of the n.

Problem 70, page 62

The formula

$$2I + B - 2$$

provides, as always, the number of primitive triangles. Figure 2.39 shows a number of different primitive triangulations, all of which must have eight small triangles inside.

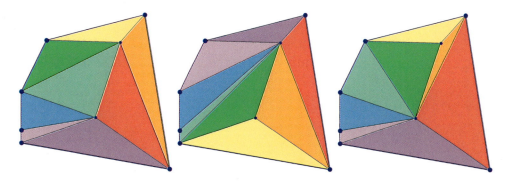

Figure 2.39: Several primitive triangulations of the polygon.

Problem 71, page 62

If you started off by considering an equilateral triangle with a horizontal or vertical base then you should have quickly dismissed that possibility (even without Pick's theorem).

Now let there be an equilateral triangle with side length a and with all three vertices in the grid. Then a^2 is an integer (use the Pythagorean theorem). What is the area of the triangle? But Pick's Rule says that all polygons with vertices in the grid have an area that is $n/2$ for some integer n. Find the contradiction[4].

Problem 72, page 62

Figure 2.40 shows the areas labeled. Most of the areas are easier to compute using familiar formulas. You might, however, have preferred Pick's formula for two of them.

---

[4]You may need to be reminded that $\sqrt{3}$ is irrational

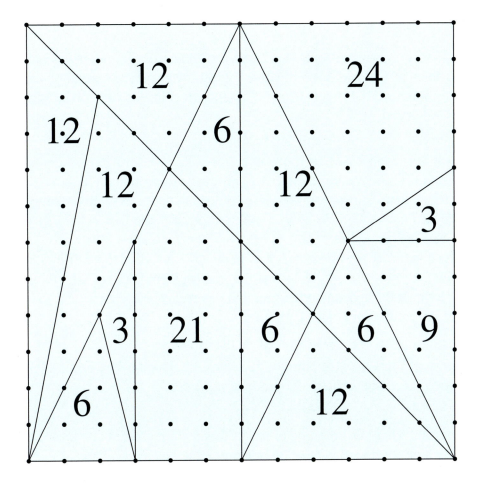

Figure 2.40: Archimedes's puzzle, called the Stomachion.

Problem 73, page 62

Each vertex lies on a the points of the grid while no other grid points lie on the surface or in the interior of the tetrahedron. J. E. Reeve (see item [8] in our bibliography) used this tetrahedron as a counterexample to show that there is no simple version of Pick's theorem in higher dimensions. This is because these tetrahedra have the same number of interior and boundary points for any value of n, but different volumes. Thus there is no possibility of a formula for the volume of a tetrahedron (or a polyhedron) that simply uses interior and boundary grid points. There are still interesting problems to address, but Pick's theorem itself does not generalize to higher dimensions as one might have hoped. Reeve's paper discusses many such related problems but it is intended for a serious mathematical audience and is not an easy read.

Problem 74, page 62

In number theory, Bézout's identity or Bézout's lemma, named after Étienne Bézout[5] states that if a and b are positive integers with greatest common divisor p, then there exist integers x and y (called Bézout numbers or Bézout coefficients) such that

$$ax + by = p.$$

Evidently we are being asked to prove only the case $p = 1$. After you have succeeded, do try to use the same method to prove the more general identity

This is not difficult to prove, if you have some knowledge of number theory and divisibility. Pick's theorem allows a different proof that relies on geometry rather than number properties.

Let a and b be relatively prime integers. In the grid, draw the line L from the origin through the point (a,b). Note that the line segment between $(0,0)$ and (a,b) does not pass through any other point on the grid.

If it did, say a different point (x,y), then

$$y/x = b/a = \text{slope of the line } L.$$

Take the point (x,y) as the grid point on the line and closest to the origin. We know that $ay = bx$ can be written as a product of primes

$$ay = bx = p_1 p_2 p_3 \ldots p_k.$$

Then, since a and b have no common prime factors, y must contain all the prime factors of b which is impossible since b is supposed to be larger.

Now, keeping in parallel to L, move the line L slowly upwards until it hits another lattice point of integer coordinates. Thus we can choose L' to be the closest parallel line to L that intersects a lattice point. Let (s, t) be the point the lattice point on L' that is closest to the origin. Consider the triangle T defined by (0, 0), (a, b) and (s, t). This triangle has no interior points and its only boundary points lie at its vertices, for if it had others then L would have hit them before it got to (s, t), which is a contradiction to how we defined (s, t). Therefore, by Pick's Theorem,

$$\text{Area}(T) = \frac{1}{2}.$$

But we have already seen in Problem 2.1.6 how to compute the area of such a triangle algebraically:

$$\text{Area}(T) = \frac{at - bs}{2}.$$

This means that

$$\frac{1}{2} = \frac{at - bs}{2}.$$

[5]Wikipedia informs us: "Étienne Bézout (1730–1783) proved this identity for polynomials. However, this statement for integers can be found already in the work of French mathematician Claude Gaspard Bachet de Méziriac (1581–1638)."

Therefore $at - bs = 1$. Substituting $c = t$ and $d = -s$ we get

$$ac + bd = 1,$$

which is what we required.

Problem 75, page 62

Yes, there must be at least one such point. One might try to find this point or show that it exists using elementary algebra, but this would get a bit messy. Much easier is to use Pick's formula for triangles.

The triangle T has base 1 and height n; from elementary geometry we know the area of T is exactly $n/2$. Since $n > 1$, the area of T must be at least 1. Using Pick's formula for triangles we see that if there were no grid points besides the vertices on or in T, the area would be only $1/2$.

We recall from Section 2.3.1 that we call such triangles primitive and a feature of our theory is that all primitive triangles have area $1/2$. In short then T, having area 1 or larger is not primitive: therefore there must be a grid point (a, b) in or on T other than one of the three vertices of T.

Chapter 3

Nim

Most of us have at one time or another played games in which we faced a single opponent: chess, checkers, monopoly, Chinese checkers, backgammon, various card games and the like. Some of these games involve *chance*. For example, the outcome of a game of monopoly depends in part on the roll of dice and on cards drawn from a stack. Most card games depend in part on which cards one draws or is dealt.

Other such games do not depend on chance: the players move alternately and each player has completely free choice of move subject only to the rules of the game. No move is dictated by the outcome of such things as rolling dice, selecting a card or spinning a dial.

In many of the games we play there are different rules for the two players (which may mean only that they use different pieces). For example in chess one side plays the white pieces and one side plays the black pieces. Games in which both sides play by precisely the same rules are said to be *impartial*.

In many games there is *imperfect information*: for most card games the players do not know what cards the opponent is holding.

In the type of games (called perfect information, impartial, combinatorial games) that we shall study there are two players, alternating moves, who see the entire positions and follow the same rules. The game ends after a finite number of moves. One such game is *Nim*.

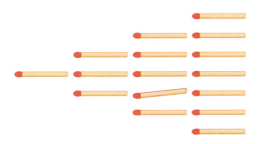

Figure 3.1: A game of Nim.

Set out matchsticks as in the figure. There are two players. Each player, in turn, removes one or more matchsticks from one of the four columns. The player removing the last matchstick wins. You can play Nim with any number of columns and any number of matchsticks in each column.

3.1   Care for a game of tic-tac-toe?

Figure 3.2: Care for a game?

Probably not. But why not? Perhaps it is because of this well-known fact.

3.1.1 (Tic-tac-toe) *Player I in a game of tic-tac-toe has a strategy that will lead in every case to either a win or a draw.*

But, in fact, that cannot be the real reason why you, as an adult, are no longer willing to play this game. The game of checkers is identical in this respect: the first player in a game of checkers has a strategy that will lead in every case to either a win or a draw. Moreover, the second player has a strategy that will in every case force a draw. Thus two completely and perfectly informed players would play every checkers game through to a draw. Every time. Just like tic-tac-toe.

The difference, however, is that no one you meet knows the strategy for checkers even though we can prove that one exists. Every schoolchild beyond a certain age knows the strategy for tic-tac-toe. Consequently tic-tac-toe retains no interest for us while checkers remains challenging and intriguing.

Prove that a strategy exists   How does one go about proving that a strategy exists without actually finding one? We shall think about this problem in the context of tic-tac-toe. Unfortunately that game is so familiar to us that it interferes with our reasoning. We adjust the rules of tic-tac-toe. The game play in *new rules tic-tac-toe* is exactly the same as before: the players alternate placing X's and O's in the squares stopping when all squares are filled or when there is a line of 3 X's or 3 O's. We consider all the end positions of the game; there are somewhat less than a hundred of these. We call some of these positions *white* positions and the rest *black*. Figure 3.2 shows an end position. We can call it black or white as we please. The winner of the game is declared following this rule: if the end position is white then player I wins, while if the end position is black then Player II wins.

An analysis of this game leads to a proof that tic-tac-toe has a strategy and we will not have to supply the strategy as part of the proof.

Problem 76 *Let an end position be defined as white if there are three X's on one of the diagonals and let every other end position be defined to be black. Show that one of the players has a winning strategy in new rules tic-tac-toe.*

Answer ☐

Problem 77 *In any new rules tic-tac-toe game prove that either player I has a winning strategy or else player II has a winning strategy.*   Answer ☐

Problem 78 *In any tic-tac-toe game (played by the ordinary rules) use Problem 77 to prove that player I has a strategy that must end in either a win or a draw.*   Answer ☐

Problem 79 (Equivalent games) *Here are the rules for the Game of 18: From a deck of cards extract nine cards numbered from 2–10 and place face up on the table. Each of two players in turn takes a card. The player wins who first obtains three cards whose sum is exactly 18. Show that this game is "the same" as a tic-tac-toe game. (This concept of two games being "equivalent" will be important later.)*   Answer ☐

Problem 80 (Simple card game) *Analyze the following card game. From a deck of cards extract the Jacks, Queens, and Kings of hearts, diamonds and spades. These nine cards are placed face up on the table. Each of two players in turn takes a card. The player wins who first obtains three cards of these types: three-of-a-kind, JQK of the same suit, or $(J\diamondsuit, Q\spadesuit, K\heartsuit)$, or $(J\heartsuit, Q\spadesuit, K\diamondsuit)$.*

Answer ☐

3.2   Combinatorial games

Mathematicians study games like tic-tac-toe, chess, checkers, and many others by describing the features that are similar. Among these similar features are that there are two players who play by certain rules known to both players, taking turns one after another, continuing until a win or a draw is declared. Both players are fully informed about the state of the game (there are no hidden elements such as cards not turned over or dice yet to be thrown). There is no element of chance. They describe such games as *combinatorial games*.

Of particular interest in any combinatorial game is whether either player can force a win and, if so, by what strategy. As we have long known, correct play by both players in tic-tac-toe must end in a draw. In 2007 it was determined,[1] after years of computer calculations, that the same is true for checkers. For chess the situation is unknown; it is possible that one side could force a win but we do not even know whether that would be white or black.

The games we shall study are all combinatorial games, but they are very special. They are said to be *impartial* in that both players must play *by the same rules* and the player who makes the last legal move is declared the winner.[2] There are no draws. For example, Tic-tac-toe (like chess and checkers) is not impartial: one player plays the X's and the other player plays the O's. The last player to make a legal move may not necessarily win (it could be a draw).

The most important impartial combinatorial game is Nim. It is the first such game to receive a complete mathematical solution. We would expect (by using the same argument as we used in Problem 77) that one of the two players in any game of this type should have a winning strategy. But how could we determine which one has the winning strategy? How could we determine what that strategy should be? How would we go about finding out the answers to these questions?

In order to motivate our development and to clarify what we are really looking for in a strategy, we shall begin with some simpler games before attacking Nim. Some of the ideas which will surface here will be central to our development.

3.2.1   Two-marker games

Two markers A and B are placed above positive integers on the number line. (Think of this as a long board with holes. The holes are numbered $1.2.3,\ldots$. Pegs marked A and B can be inserted in the holes.) For example, we might have placed A at hole number 4 and B at hole number 9 as indicated in Figure 3.3.

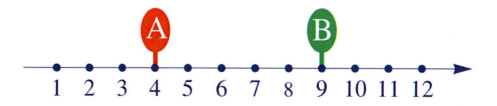

Figure 3.3: A game with two markers at 4 and 9.

The two players move alternately. A move consists of moving either one of the markers to the left as far as one wants with the proviso that B stays to the right of A. (Markers must be placed above an integer.) The player who makes the last legal move wins.

---

[1] Jonathan Schaeffer et al., *Checkers is solved*, Science Vol. 317 no. 5844, pp. 1518–1522.

[2] This is called the *normal play condition*. We will also, much later in the chapter, consider a different kind of combinatorial game where the last player to make a legal move loses.

Figure 3.4: The ending position in a game with two markers.

Figure 3.4 illustrates the end position in any game of 2–markers. This must occur only when A is at position 1 and B is at position 2. To win in a game of 2-markers you would be well-advised to keep the end position always in mind.

Example 3.2.1 Let us play the two-marker game with markers at 4 and 9 (as in Figure 3.3). The player whose turn it is has seven possible moves: he or she can move A to any one of the positions 1, 2 or 3 or can move B to any of the positions 5, 6, 7 or 8. The game ends when a player has no move available. This must occur only when A is at position 1 and B is at position 2. The player who made the final legal move wins.

A bit of reflection shows that, for this game, we can guarantee a win by the following procedure. We move B to 5. Now, according to the rules, our opponent cannot move B. He or she must move A. Whatever move our opponent makes, we answer by moving B right next to A. Following this procedure, we see that eventually our opponent must move A to 1 and we answer by moving B to 2. We won. ◀

The strategy in the example would work no matter what the original position was, as long as it was possible for us to move B at our first turn. For two-marker games we can say that there are precisely *two kinds of positions*: ones in which we can make a good move and ones in which no good move can be made.

3.2.2 Three-marker games

Let us complicate the game by introducing a third marker C on an integer to the right of A and B. For example, we might start with the position indicated in Figure 3.5 with markers at 4, 9, and 12.

The rules are as before. When it is our turn, we may move any of the three markers as far as we wish to the left as long as the relative order of the marker from left to right remains the same—B must stay between A and C. The game ends when a player has no move available.

Figure 3.6 illustrates the end position in any game of 3–markers. This must occur only when A is at position 1, B is at position 2, and C is at position 3. Keep this end position in mind as your final goal.

Figure 3.5: A game with three markers at 4, 9, and 12.

Figure 3.6: The ending position in a game with three markers.

Problem 81 *Find a strategy for the three-marker game. Begin by experimenting with a marker board and the three markers A, B and C.*

Answer □

3.2.3 Strategies?

Let us digress for a moment and consider a game like chess or checkers. What is it that distinguishes a strong chess player from a weak one? Obviously, that is not a question which can be answered easily—a strong player knows the openings, has studied many combinations, knows the endings, and can look ahead many moves.

But there is one feature we can focus on which will be central to our development of the marker games and Nim. A good chess player will recognize many positions as desirable to achieve. For example, very early in one's learning of the game of chess, one realizes that if one can achieve a position in which one has the king and the queen while the opponent has only the king, the game can be won quickly.

As one improves, one recognizes more and more of such desirable positions. Thus, the good chess player can have many, many subgoals when playing chess. He does not have to see how to checkmate the opponent from the very beginning of the game—he must only try to achieve one of these many desirable positions. The same is true of checkers and of other of these two-person games of skill. Our device for discovering strategies for Nim and the marker games is to find a way of determining all of these sub-goals for the given game.

Perfect strategies? Usually, by a strategy, we mean some method we can use to improve our chances of winning. In Example 3.2.1 we did not merely improve our chances of winning. We can win *every time* provided we start first. And we will lose every time that we start second if we are playing against an informed opponent.

There are many strategic advantages that a clever and informed chess player can use. An international grandmaster will win any game he or she plays against a lesser player. But there is as yet no known perfect strategy for chess. For the games in this chapter we are not content with just strategies. We want perfect strategies.

3.2.4 Formal strategy for the two-marker game

Let us formalize what we discovered in the two-marker game. We use the suggestive notation $B = A + 1$ to describe the fact that B is adjacent to A. Observe three facts:

1. $2 = 1 + 1$, thus the final position satisfies the equation $B = A + 1$.

2. If $B = A + 1$, then any move whatsoever results in a position for which that equation is no longer satisfied.

3. If $B \neq A + 1$, there is a move which results in the equation being satisfied.

Let us say that a position is *balanced* if it satisfies the equation $B = A + 1$ and that it is *unbalanced* if $B \neq A + 1$. With this language, the three observations above become:

3.2.2 (Balanced positions in marker games) *Every position in the game is either balanced or unbalanced.*

1. *The final position is balanced.*

2. *If a position is balanced, then any move whatsoever results in a position that is unbalanced.*

3. *If a position is unbalanced, there is a move which results in a position that is balanced.*

Once we have articulated the situation in this balanced and unbalanced language we can easily prove that we do in fact have a strategy. We can always move from an unbalanced position to a balanced position. Our opponent always receives a balanced position and must destroy the balance. The final position is balanced. Eventually, after some finite number of moves, our opponent is faced with this final balanced position and has no move. We win!

The *balanced positions* are the subgoals that we seek. This will be true in *all* the games we shall look at in this chapter. It is very important that one understands the notion of *balanced* positions before proceeding further, so we suggest that you reread the preceding discussion and relate it to the two-marker game before going on.

3.2.5   Formal strategy for the three-marker game

The three-marker game has been used by some teachers in elementary school as a device for motivating children to practice addition and subtraction. The children usually discover, through playing the game repeatedly, that the balanced positions are given by the equation

$$A + B = C.$$

They do not know anything about balanced positions, of course; they just discover that they can win if they can obtain those positions.

Following the strategy that the children discovered, let us say that a position in the three-marker game is *balanced* if it satisfies the equation $A + B = C$ and that it is *unbalanced* if $A + B \neq C$.

Problem 82 *Verify that each of the three parts of Statement 3.2.2 apply to the three-marker game.*   Answer □

Problem 83 *Discover the balancing positions for the four-marker game and prove that the same three rules apply to them.*   Answer □

Problem 84 *What are the balancing positions for the five and six-marker games?*
                                                                    Answer □

3.2.6   Balanced and unbalanced positions

Generally we are seeing that, in games of this type, an analysis using the ideas of balanced and unbalanced positions[3] leads to a strategic way of thinking about the game. Any end position is balanced. A balanced position always leads to an unbalanced position. An unbalanced position always allows a move to some balanced position. If this is so the strategy is clear.

It might seem that determining which positions in a game are balanced and which are unbalanced takes some considerable skill. It was easy for the two-marker game, rather harder for the three and four-marker games, and apparently formidable for a five-marker game. In fact, though, it need not take skill, but it does take patience. We can do this formally for *any* game of the kind we study in this chapter.

We assume, as always, that the players alternate turns at making moves according to the same rules. After a finite number of moves the game ends and the last person to move is declared the winner.

---

[3]In the literature the *unbalanced* positions are often called *N-positions*, (because the *next* player is to win), while the *balanced* positions are known as *P-positions* (because the *previous* player is to win).

Defining balance You might have noticed that in trying to determine the balanced positions in a game, there is a sense of "working backwards." The final position is balanced. We then seek simple balanced positions that lead quickly to the final position. (Think of our discussion about chess.) Then we seek positions that lead to one of the positions we have already determined to be balanced.

We can put these ideas into a formal setting. This material is somewhat abstract but not difficult. We merely define carefully what we mean by *starting at the end of the game* and what we mean when we say a position is *balanced or unbalanced*. The definition rests on the principle of mathematical induction.

A formal way of presenting these ideas and checking the accuracy of our intuitions is to introduce a *balancing number* for any position in a game. If G is one of our games and p is a position in that game we define Balance(p) by these rules:

1. If p is an end position in the game G then Balance(p) $= 0$.

2. If p is not an end position in the game G, first find all the positions $p_1, p_2, \ldots, p_n$ that could be obtained from the position p in one legal move. We use the notation

$$p \rightsquigarrow p_1, p_2, \ldots, p_n$$

   to indicate that any move that can follow p is in this list. Then compute the list of numbers

   $$\text{Balance}(p_1), \text{Balance}(p_2), \ldots, \text{Balance}(p_n).$$

3. Balance(p) is defined to be zero if zero does *not* appear in the list and to be 1 if zero *does appear* in the list.

A position with a balancing number of zero is said to be *balanced*. If the balancing number is 1 then it is *unbalanced*.

Note how these rules will always require a position with a zero balancing number to lead to nothing but positions with a balancing number of 1. Observe too that these rules will always require a position with a balancing number of 1 to lead to at least one position with a balancing number of 0. Our definition is designed precisely around the rules that we devised in Statement 3.2.2 for our marker game.

This is an example of a recursive definition; we have to build up the values of the function Balance(p) step by step starting close to end of the game. In a way we would have to play the game backwards.

The way that we have defined the balancing number shows that

- Any end position has a balancing number of zero.

- If a position has a balancing number of zero then all positions which follow it in the game have a balancing number of one.

- If a position has a balancing number of one then there is at least one position that follows it in the game that has a balancing number of zero.

Thus balanced and unbalanced positions are defined now in any game, and they behave precisely as we required for the marker games in Statement 3.2.2. The strategy in any game is the same: always (if you can) leave your oppenent a balanced postition, forcing him to unbalance it at his next move. Since the game ends in a finite number of steps at a final balanced position, the player who can follow this strategy must have made the last move and is declared the winner.

Depth of a position   In practise it is easy to see that this recursive definition will assign a value to each position in any game. To make it more precise how this is done let us introduce the notion of *depth* of a position. This is just a measure of the maximum number of moves left in the game. Any end position (there may be several) has no further moves possible and is said to be at *depth zero*. Such positions are always balanced. If a position can move only to a depth zero position, then it is said to be at *depth one*. Such positions are always unbalanced.

If a position that is not at depth zero or depth one can move only to a depth zero or a depth one position, then it is said to be at *depth two*. Such positions may be balanced or unbalanced. We would have to check. Generally a position that is not itself at depth 0, 1, 2, ..., or $n-1$ and that can move only to such a position, is said to be at depth n. At depth n the game must end in at most n moves.

All games solved!   By this simple definition we have precisely defined, for any game of this type, how a position may be considered balanced or unbalanced and we have a method for computing that fact. Thus we can solve all games!

Well not all games, because not all games are of this type. Tic-tac-toe, chess, and checkers have rules that are different for the two players (e.g., one player plays the X's and the other the O's). The game may end in a draw. The rule is not that the last player who is able to make a move wins.

But, for finite games of the last-move type discussed so far, the solution is exactly this. Compute all balanced positions and play the game in such way (if possible) as to leave your opponent only balanced positions. If you start play with an unbalanced position then you will surely win. If you start play with a balanced position then, provided your opponent makes just one mistake, you will win.

Is this practical?   Recursive definitions like this one, however, are particularly difficult and tedious to compute. On the other hand they are particularly easy to program and run on a computer. Unless the game has billions and billions of possible positions (like chess and checkers do), a short amount of time will enable a full computation of all the balanced positions. A human computation by hand could be extremely slow and tedious.

The moral is do not play any of these games against a computer; you will surely lose. It may be safe to play against a human, unless she has figured out a cleverer way to find balanced positions without having to compute Balance(p) for all positions in the game in the way the recursive definition prescribes.

For us the problem now is not finding all balanced positions, but finding some elegant and simple way of describing them without having to resort to brute force and compute Balance(p) for every position in the game.

Problem 85 *In the game of 2–pile Nim, players in turn take matchsticks (one or more) from one of two piles. The player to take the last matchstick wins. Compute the depth and balancing numbers for enough positions that you can make a reasonable conjecture about which positions are balanced and which are unbalanced.* Answer □

Problem 86 (Red and black argument) *Suppose that all the positions in a game are described as either red or black and that these three statements are true:*

1. *Any end position is red.*

2. *Any red position can move only to a black.*

3. *From any black position there is at least one move to a red position.*

Show that the red postitions are balanced and that the black postitions are unbalanced.

Answer □

Problem 87 *In Problem 85 you would have made a conjecture about the balanced and unbalanced positions in the game of 2–pile Nim. Use the red and black argument to prove this conjecture.* Answer □

Problem 88 *The game of 2–pile SNIM is played exactly as Nim but each player has the option of adding one matchstick to a pile or removing as many as he pleases from that pile. Show that, even though the balanced positions are the same as for Nim, there is no winning strategy. What is wrong here?* Answer □

Problem 89 *In our four-marker game (in the answer to Problem 83) we said that a position was balanced if and only if the equation $D - C = B - A$ was satisfied. Use the red and black argument to prove this fact.* Answer □

Problem 90 *In a game every move from a balanced position will produce an unbalanced position. In some games the reverse is also true: every move from an unbalanced position will produce a balanced position. How would you describe those games?*

Answer □

Problem 91 *If Player I faces an unbalanced position the challenge for him is to select a correct move (there must be at least one) that rebalances and leaves a balanced position. If Player II faces a balanced position then every move she makes will (unfortunately) produce an unbalanced position. Is there any strategic choice for Player II in such a game?*   Answer □

3.2.7   Balanced positions in subtraction games

The analysis of the balanced and unbalanced positions in the two-marker and three-marker games presented little difficulty. The four-marker game was a bit tougher, and the five and six-marker games of Problem 84 may well have defeated you.

For a little more practice with these ideas here are some simpler games where the balanced and unbalanced positions are in some cases easy to work out. Remember that every position must be either balanced or unbalanced: we are looking for a fast and easy way of finding out which is the case for any position.

Problem 92 *In this game there is one pile of matchsticks and each player removes 1, 2, 3, or 4 sticks at a time. The winner is the one removing the last matchstick. What are the balanced positions for this game?*   Answer □

Problem 93 *In this game there is one pile of matchsticks and each player removes 1, 4, 9, 16, ... sticks at a time, always restricted to a perfect square. The winner is the one removing the last matchstick. Find all the balanced positions less than 25 for this game?*   Answer □

Problem 94 *Find all the balanced positions between 25 and 100 for the game of Problem 93.*   Answer □

Problem 95 *Do you have a conjecture as to a formula that will produce all the balanced positions for the game of Problem 93.*   Answer □

Problem 96 *In the most general one-heap subtraction game there is one pile of matchsticks and each player removes an allowed number of sticks at a time, always restricted to numbers from a given subtraction set S. The winner is the one removing the last matchstick. Thus Problem 92 is a one-heap subtraction game with $S = \{1,2,3,4\}$. Problem 93 is a one-heap subtraction game with $S = \{1,4,9,16,25,36,...\}$. Find the balanced positions for a subtraction game given the subtraction set*

$$S = \{1,2,3,4,5,6,7,8,9,10\}.$$

Try to experiment with other choices of S.   Answer □

Problem 97 *Give rules for a two-heap subtraction game and find some balanced positions in the simplest cases.*   □

Remarks For some of the one-pile subtraction games the analysis is fairly easy. But, even when things prove difficult to compute, the resolution always follows from our balanced and unbalanced accounting. For the marker games the same is true. By the time we get to five and six-marker games (as in Problem 84) we ran into considerable trouble finding the balanced positions. Equations defining balanced positions similar to those for the two, three and four-marker games did not come to mind readily. There is a reason for this and we will discover that reason later. Instead of pursuing the marker and subtraction games further at this time, we will continue with some other games. But we will return to the marker games later.

3.3 Game of binary bits

The game of binary bits that we introduce in this section contains much of the important structure of all of our games and is fundamental to all of combinatorial game theory. We start with an equivalent game that provides our introduction to the bits game.

3.3.1 A coin game

This game is played with coins—pennies, nickels, dimes, and quarters. Each position in the game is n piles of 0–4 coins such that each pile contains at most one coin of each type. The rules of the game are

1. Each play of the game requires a player to remove one or more coins from one of the piles.

2. Optionally the player may also add one or more coins to the same pile provided the coins added in are of lower value than the highest-value coin removed. (E.g., a player removes a dime and a penny and can add a nickel (if there is not one there already) but cannot add a quarter.)

3. The player to take the last coin is the winner of the game.

The easiest way to display a position in the game, both for the purposes of writing about it and for the purposes of play itself, is to arrange the coins in a rectangular display of 4 rows and n columns as in Figure 3.7. Pennies are recorded on the bottom row, nickels on the row above and so on until the quarters are displayed on the top row as Figure 3.7 illustrates.

We do not yet see what positions in such a game would be balanced or unbalanced, but a person aware of the strategy would see immediately that the position in Figure 3.7 is unbalanced. A balancing move is to take a dime from the 5th pile and toss in a nickel. That takes only a couple of seconds to compute

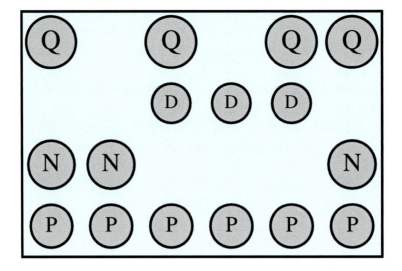

Figure 3.7: Position in the coin game.

if one knows the strategy. Moreover a strong player will notice that there are exactly two other balancing moves that would have worked too. Did you?

Problem 98 *Play some simple coins games with one, two or three piles. What did you observe?*   Answer □

Problem 99 *Show that the coin game must end in a finite number of moves.*
                                                           Answer □

3.3.2 A better way of looking at the coin game

In analyzing this game one soon realizes that the notations ⓟ, ⓝ, ⓓ, and ⓠ are completely unnecessary since the position in the rectangular array already determines which coins appear. That means we need record only YES or NO in each case.

The traditional way to do this now, especially since the advent of computers, would be to use binary bits—the bit 1 is used for YES and the bit 0 is used for NO. That means that Figure 3.7 can be written out instead using the simpler Figure 3.8.

Also we can simplify the moves in the game if we realize that removing a coin simply changes a YES to a NO, i.e., it changes a 1 bit to a 0 bit. Similarly adding a coin changes a NO to a YES, i.e., it changes a 0 bit to a 1 bit. We are just flipping bits, which is a good description of what computers do. Thus, if we translate the coin game to binary bits, we arrive at the binary bits game of Section 3.3.3 which is exactly identical to it.

Problem 100 (A card game) *In this game a deck of cards is shuffled and four-teen cards are dealt on the table face up. A play in the game requires the player*

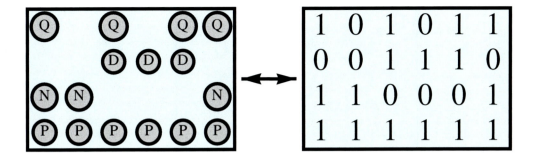

Figure 3.8: The same position in the coin game with binary bits.

to remove one of the cards. He then has the option of removing more of the cards of the same suit that have lesser face value, and/or adding (from the left-over pile) any cards of the same suit that have lesser face value. How would you analyze this game? Answer □

3.3.3 Binary bits game

In the game of *binary bits* we start off with a $m \times n$ rectangular array of zeros and ones. There are m rows and n columns and only the numbers 0 and 1 can appear. As is often the case, the numbers are called *bits*. A legal move of the game is described this way:

1. The player selects a 1 bit in some position and changes it to a 0 bit.

2. The player may optionally change any or all of the bits in the column below the selected bit 1.

Play evidently stops when all the bits have been changed to 0. The player who made the last legal move wins.

At first it seems obvious that the game eventually stops. A moment's reflection, however, may give us pause. As the game progresses some moves may add 1 bits, so the total count of 1 bits does not always go down. In Problem 101 you are asked to show that the game is finite. This, we recall, is essential if our analysis in terms of balanced and unbalanced positions is to be successful.

Example 3.3.1 A move in a 5×3 game is illustrated in Figure 3.9 Here the player elected to change one of the 1 bits in the second column, and he also flipped two of the lower bits.

Can you spot whether this was a good move? Was there a better move? ◀

The $m \times 1$ game · Here there is but one column and the strategy should be obvious. The player to start simply chooses the topmost 1 bit and changes that bit and all the ones below it to zero bits. The game is over and he wins. A position with any 1 bits is unbalanced.

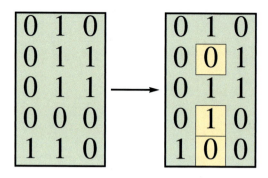

Figure 3.9: A move in a 5×3 game of binary bits.

The $m \times 2$ game   Here there are two columns and the strategy is obvious ...after some thought. In Problem 102 you are asked to solve the game. The strategy that works is called *the mirror strategy* and plays an important role in game theory.

The $m \times 3$ game   Here there are three columns and the strategy is no longer obvious at all. At this point the game becomes rather more interesting. We know that an analysis of balanced and unbalanced positions will result in a completely solved game but we do not yet know how to do that in any simple way.

Problem 101 *Show that every game of binary bits must end in a finite number of steps.*   Answer □

Problem 102 *Find a complete strategy for the $m \times 2$ game of binary bits.*   Answer □

Problem 103 *Which, if any, of the positions in the 5×3 games of Figure 3.10 are balanced?*   Answer □

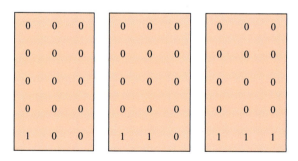

Figure 3.10: Which positions are balanced?

Problem 104 *Which, if any, of the positions in the 5×3 games of Figure 3.11 are balanced?*   Answer □

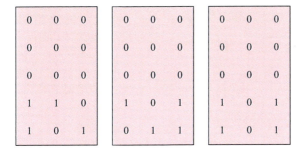

Figure 3.11: Which positions are balanced?

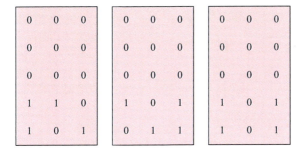

Figure 3.12: Which positions are balanced?

Problem 105 *Which, if any, of the positions in the 5 × 3 games of Figure 3.12 are balanced?* Answer □

Problem 106 *Do you have a conjecture?* Answer □

Problem 107 *Define a position in a m × 3 game to be even if there are an even number of 1 bits in each row. Define a position in a m × 3 game to be odd if there is at least one row containing an odd number of 1 bits. Check each of the following:*

1. *The end position of the game is even.*

2. *If a player makes a move from an even position it will surely result in an odd position.*

3. *If a player faces an odd position there is always a choice of move that leaves an even position.*

Answer □

Problem 108 *Give a complete solution for the m × 3 game of binary bits.*

Answer □

Problem 109 *Are you prepared to announce a solution for the m × n game of binary bits?* Answer □

Problem 110 *Describe all the balancing moves in the coin game displayed in Figure 3.7.*   Answer □

Problem 111 *In the coin game one can change the rules to allow more coins in each pile. For example:*

1. *Each play of the game requires a player to remove all the coins of the same type from one of the piles.*

2. *Optionally the player may also add coins to or subtract coins from the same pile provided the coins added or subtracted are of lower value than the coins initially removed. (E.g., a player removes all dimes and then can add or subtract as many pennies and nickels as he pleases, but cannot add any quarters.)*

3. *The player to take the last coin is the winner of the game.*

How does this change the game?   Answer □

Problem 112 *In the coin game one can change the rules to allow any player to keep the coins that he has removed. How does this change the game?*

Answer □

Problem 113 (A number game) *A game similar to binary bits starts with a $m \times n$ rectangular array of arbitrary numbers. A legal move of the game is to change any nonzero number to zero and, optionally, change any or all of the numbers in the column below the selected number. The last player to move wins. Analyze this game.*   Answer □

Problem 114 (A word game) *This word game is also similar to the game of binary bits. The players start with three or more words. A player moves in this game by selecting a word and a letter that appears in that word. He must remove all appearances of that letter in the word chosen and may, optionally, add in or remove any other letters that are earlier in the alphabet. For example if the six words are*

[*Twas brillig and the slithy toves*]

*then a legal move would be to select the "l" in **brillig** and remove both of them. The "r" cannot be removed but the other letters can and any letters a—k could be added in, for example*

brillig ⤳ abbrek

would be allowed. The last player to move wins. Analyze this game.

Answer □

Problem 115 *Are you prepared to announce a solution for the game of Nim?*

Answer □

3.4 Nim

The classical game of Nim is played as follows. Four piles of matchsticks (or cards or coins) containing 1, 3, 5, and 7 sticks respectively, are placed on a table as indicated in the diagram.

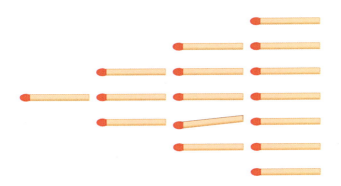

Figure 3.13: A game of Nim.

One of the players removes one or more sticks (as many as he likes) all from the same pile. Then the opponent does the same from the remaining sticks. *The player who takes the last stick wins.*

We do not need match sticks to play the game, of course. We could instead consider the quadruple of numbers $(1, 3, 5, 7)$ and lower one of those numbers to start the game. Then our opponent would lower one of the remaining numbers. The game ends when there are no positive numbers left; that is, when all four numbers are zero. This corresponds to having no matchsticks left on the table.

The general Nim game There is nothing special about the numbers $(1, 3, 5, 7)$. We can play the same game with any number of piles of matchsticks, and the number of sticks in each pile can be chosen as we like. We can describe a Nim game with k piles containing $n_1, n_2, \ldots, n_k$ sticks in the piles by writing

$$(n_1, n_2, \ldots, n_k).$$

Our objective is to find a winning strategy that will apply to *every* Nim game, no matter how many piles there are and no matter how many sticks are in each pile.

3.4.1 The mathematical theory of Nim

At this stage we know how to solve the game of Nim in a technical sense. We can simply describe all balanced and all unbalanced positions. Lacking any better ideas we just start at the end of the game working backwards. Maybe the real structure will emerge. Or perhaps the real structure will remain mysterious even after seeing all the balanced positions.

The complete mathematical theory for the game of Nim was discovered by Charles L. Bouton who published his observations in the research journal *Annals of Mathematics* in 1901. Bouton's paper marked the beginning of the theory of what are called combinatorial games.

Research papers announce and publish results in a relatively compact form for the mathematical community. His complete paper is reproduced in our appendix on page 239. Our interest here is not in the result or formal proofs of the result, but in *discovery*.

If you are impatient to learn the strategy and beat all your friends at Nim go to the appendix and read Bouton's paper. If you wish, as we do, to go through the process of discovering a clever mathematical theory, begin instead by playing the game and looking for the underlying structures.

3.4.2 2–pile Nim

As an easy warm-up, let us begin with 2–pile Nim. Here we have two piles of sticks and our objective is *to take the last stick*. Our analysis will use the usual ideas of balanced and unbalanced positions.

A position in two-pile Nim is described by a pair of numbers (m, n) representing two piles of sticks, one containing m sticks and the other containing n sticks.

Starting at the bottom, we know that $(0,0)$ is balanced and can deduce that therefore $(1,0)$ and $(0,1)$ are unbalanced. Carry on until you are able to spot the pattern.

Problem 116 *Discover the strategy for 1–pile Nim.*   Answer □

Problem 117 *Discover the strategy for 2–pile Nim.*   Answer □

Problem 118 *The mirror strategy that we used in Problem 117 works for 2–pile Nim, but is not much help with 3–pile Nim or 4–pile Nim. Even so, there are some situations where it can work. Show that a mirror strategy will win a game of 4–pile Nim if the opening position is of the form (m, n, m, n).*   Answer □

Problem 119 (Kayles) *The mirror strategy works for a number of other games. Try it on the game of Kayles. Line up a number of coins in a row so that each coin touches its neighbors as in Figure 3.14.*

Figure 3.14: Coins set up for a game of Kayles.

The rules of the game are that a player may remove a single coin or two coins that touch each other. The last player to move wins. Show how a mirror

(Tweedledum-Tweedledee) strategy can be used to solve this game.

Answer □

Problem 120 (Circular Kayles) *Find a strategy for the game of Kayles when the coins are arranged in a circle instead of a straight line.*   Answer □

3.4.3   3–pile Nim

Let us proceed to 3–pile Nim. Here things are quite a bit more complicated. The game has a rather complex structure and it will take a while to discuss the balanced positions.

For 2–pile Nim we discovered in Problem 117 that the balanced positions are *those which have the same number* in each pile. These are games of the form (n,n), indicating two piles so that each pile has n sticks. Unfortunately knowing the complete strategy for 2-pile Nim doesn't give us any clues as to the strategy for 3-pile Nim.

We also could have expressed our solution of 2–pile Nim in terms of the mirror strategy. Again this doesn't help us in finding a solution to 3–pile Nim. The only method we have that is general enough to lead us in the right direction is to search for balanced and unbalanced positions.

You may wish to find an opponent with whom to play a few games, just to get a feeling for the game. Start with games which do not have many sticks in each pile.

The situation is a bit like chess or checkers. By playing a few games, one can learn how to play better, but to become a really good player, one must also begin to learn something about the structure of the game. A difference is that in order to become an excellent chess player, we must devote a great deal of time to the subject. And no one knows a perfect chess strategy. With Nim we shall eventually see what the perfect strategy is. And it involves only a few ideas.

Starting off at the bottom we can easily construct a few balanced positions. As usual $(0,0,0)$ is balanced and that shows us that $(1,0,0)$, $(0,1,0)$, and $(0,0,1)$ are unbalanced. At the next level $(1,1,0)$, $(0,1,1)$, and $(1,0,1)$ are balanced so that $(1,1,1)$ must be unbalanced. Carry on. Does a pattern emerge?

Problem 121 *Are the two positions $(1,2,2)$ and $(1,1,2)$ balanced or unbalanced?*

Answer □

Problem 122 (The position $(1,2,3)$ is balanced) *Go through all the details necessary to check that $(1,2,3)$ is balanced.*

Answer □

3.4.4   More three-pile experiments

In solving a number of our problems we took advantage of the fact that we knew all the balanced positions in two-pile Nim. Thus we can easily spot whether $(m,n,0)$ is balanced or not because this is identical to playing the game (m,n) in two-pile Nim. While this was a bit of help, it proves to be a dead end for finding the pattern that describes the three pile game.

This is disappointing since it means a familiar technique is not going to work. Going from two-pile Nim to three-pile Nim presents us with a different game. Mastery of the former gives us only minimal assistance in playing the latter. The same will happen with 4-pile Nim: even if we compile a list of all balanced positions in 1-pile, 2-pile, and 3-pile Nim, we will still have trouble.

We need to find a new kind of pattern. If you have experimented with a number of small games, you have undoubtedly begun to pick up certain patterns although, at this stage, it is still not clear how to exploit those patterns.

Example 3.4.1 Did you notice that the games

$$(1,0,1),\ (2,0,2),\ (2,1,3),\ (3,0,3),\ \text{and}\ (3,1,2)$$

are all balanced? Compare that with the fact that the games

$$(1,1,2),\ (2,2,4),\ (2,3,5),\ (3,2,5)\ \text{and}\ (3,3,4)$$

are all unbalanced.

These two groups of games form a certain pattern obtained from the first in each list. Thus, $(2,0,2)$ is the *double* of $(1,0,1)$. The other three games in the first group can be obtained by a *near-doubling* from $(1,0,1)$ or $(2,0,2)$ by adding a stick to exactly two of the piles.   ◄

Examples such as these might suggest that doubling or near-doubling a game does not change its status – if the original game is balanced, so are the resulting games obtained by doubling or near doubling. If that turns out to be true, it will give us a large collection of positions whose status we will know.

What about the other near doubles we obtain by doubling the number of sticks in each pile and adding one stick to exactly one of the resulting piles? Or to all three of the resulting piles? For example, from the game $(1,0,1)$ we would get the games $(2,1,2)$, $(3,0,2)$, $(2,0,3)$, and $(3,1,3)$. What do you think happens? Work it out, make a conjecture, and then see if your conjecture is valid for the examples in Problems 123–126.

Problem 123 *Which of the games*

$$(2,4,6),\ (2,5,7),\ (3,4,7),\ \text{and}\ (3,5,6)$$

are balanced and which are unbalanced?   Answer □

Problem 124 *Which of the games*

$$(2,6,8),\ (2,7,9),\ (3,6,9),\ \text{and}\ (3,7,8)$$

are balanced and which are unbalanced?   Answer ☐

Problem 125 *Which of the games*

$$(3,4,6), \ (2,5,6), \ (2,4,7), \ \text{and} \ (3,5,7)$$

are balanced and which are unbalanced?   Answer ☐

Problem 126 *Which of the games*

$$(3,6,8), \ (3,7,8), \ (3,6,9), \ \text{and} \ (3,7,9)$$

are balanced and which are unbalanced?   Answer ☐

Problem 127 *Study the patterns of Problems 123–126. How are they related to the games* $(1,2,3)$ *and* $(1,3,4)$*? Do you see any connection between the strategies for these games,* $(1,2,3)$ *and* $(1,3,4)$*, and the games in Problems 123–126.*
                                                                  Answer ☐

3.4.5 The near-doubling argument

> *Can we yet spot the structure of the balanced positions in 3–pile Nim? A flash of insight would help and perhaps you have had one. If not, then the line of reasoning we now follow will lead us closer to the moment of recognition.*

The bright idea we need to progress further is apparent in the experiments we have so far performed, provided we look at things from a new point of view. We noticed that all these positions were balanced:

$(1,1,0)$ as well as $(2,2,0)$, $(1+2,1+2,0)$, $(2,1+2,1+0)$, and $(1+2,2,1+0)$.
This includes $(3,2,1)$, so $(1,2,3)$ would be balanced too.

The same kind of doubling produces yet more balanced positions:

$(1,2,3)$ as well as $(2,4,6)$, $(1+2,1+4,6)$, $(2,1+4,1+6)$ and $(1+2,4,1+6)$.

Starting at $(2,2,0)$ and using the same pattern produces

$(2,2,0)$ as well as $(4,4,0)$, $(1+4,1+4,0)$, $(4,1+4,1+0)$ and $(1+4,4,1+0)$.

A little checking shows that these too are balanced.

If we were to include in this list all the different permutations we would recognize that we have obtained all of the balanced positions close to the end of a 3–pile game just by doubling and redoubling $(1,1,0)$ and maybe adding a couple of 1's each time. If we continue this process further perhaps we can generate all balanced postions.

Examples such as these suggest that doubling a balanced game does not change its status. Nor does doubling and adding 1 to two of the piles. If that turns out to be true, it will give us a large collection of positions whose status we will know. Here is our conjecture.

Near doubling   Start with any Nim position (x,y,z). Any of the four positions

$$(2x,2y,2z),\ (2x+1,2y+1,2z),\ (2x+1,2y,2z+1),\ \text{or}\ (2x,2y+1,2z+1)$$

are said to be *near-doubles* of (x,y,z). Note that a position cannot be a near double of more than one choice of (x,y,z).

3.4.2 (Near doubling argument) *A position (x,y,z) in 3–pile Nim is balanced if and only if it is a near-double of another balanced position.*

To prove this statement we use an argument that should be familiar to us. We used it before in our even/odd analysis of the game of binary bits. Let us call a position a *red position* if it is near-double of a balanced position. Every other position is said to be a *black position*.

The end position is red   The end position in a three pile game of Nim is $(0,0,0)$. Since this is balanced and is its own near-double the end position is red.

Any red position must move only to a black   Start with any of these red positions:

$$(2x,2y,2z),\ (2x+1,2y+1,2z),\ (2x+1,2y,2z+1),\ \text{or}\ (2x,2y+1,2z+1)$$

where we are assuming that (x,y,z) is balanced.

It is enough for our argument to consider only moves that take away sticks from the first pile—the argument is the same for the other cases. Taking away an even number $2k$ of sticks from the first pile results in

$$(2x,2y,2z) \rightsquigarrow (2[x-k],2y,2z),$$
$$(2x+1,2y+1,2z) \rightsquigarrow (2[x-k]+1,2y+1,2z),$$
$$(2x+1,2y,2z+1) \rightsquigarrow (2[x-k]+1,2y,2z+1)$$

and

$$(2x,2y+1,2z+1) \rightsquigarrow (2[x-k],2y+1,2z+1).$$

We recognize a doubling or near-doubling of the pile $(x-k,y,z)$. But $(x-k,y,z)$ must be unbalanced since it came from a move out of the balanced position (x,y,z). Consequently all of the resulting positions are black, i.e., all of our red positions have moved to black if we remove an even number of sticks.

Start again with any of these red positions:

$$(2x,2y,2z),\ (2x+1,2y+1,2z),\ (2x+1,2y,2z+1),\ \text{or}\ (2x,2y+1,2z+1)$$

but this time remove an odd number $2k-1$ of sticks, again from the first pile:

$$(2x,2y,2z) \rightsquigarrow (2[x-k]+1,2y,2z),$$
$$(2x+1,2y+1,2z) \rightsquigarrow (2[x-k+1],2y+1,2z),$$
$$(2x+1,2y,2z+1) \rightsquigarrow (2[x-k]+2,2y,2z+1)$$

and

$$(2x, 2y+1, 2z+1) \rightsquigarrow (2[x-k]+1, 2y+1, 2z+1).$$

Again we recognize all of these positions to be black, i.e., all of our red positions have moved to black if we remove an odd number of sticks.

Any black position can be moved to at least one red We need to consider several cases of black positions and, for each one, determine how to make the correct move to a red position.

1. Suppose that $(2x, 2y, 2z)$ is a black position. Then (x, y, z) is unbalanced and so there is a balancing move which leaves, say, the position $(x-k, y, z)$. Since that position is balanced, the doubled position

$$(2x-2k, 2y, 2z)$$

is a red position. This gives us a way to move from the black to the red if we start at $(2x, 2y, 2z)$ assumed to be a black position. Take away $2k$ sticks.

2. Suppose that $(2x, 2y+1, 2z+1)$ is a black position. Then (x, y, z) is unbalanced and so there is a balancing move which leaves, say, [Case 2a] the position $(x-k, y, z)$ or [Case 2b] the position $(x, y-k, z)$ or [Case 2c] the position $(x, y, z-k)$. We need consider only the first two cases.

In Case 2a the position $(x-k, y, z)$ is balanced, hence the near-doubled position

$$(2[x-k], 2y+1, 2z+1)$$

is a red position. This gives us a way to move from the black to the red if we start at $(2x, 2y+1, 2z+1)$ assumed to be a black position. We remove $2k$ sticks from the first pile.

In Case 2b the position $(x, y-k, z)$ is balanced, hence the near-doubled position

$$(2x, 2[y-k]+1, 2z+1$$

is a red position. This gives us a way to move from the black to the red if we start at $(2x, 2y+1, 2z+1)$ assumed to be a black position. We remove $2k-1$ sticks from the second pile.

3. Suppose the starting position is $(2x+1, 2y, 2z)$; this is always a black position since two of the entries are even. If (x, y, z) is balanced then there is an obvious move: take away 1 from the first pile to produce the red position $(2x, 2y, 2z)$. If, however, (x, y, z) is unbalanced we can balance it to the position $(x-k, y, z)$ or perhaps $(x, y-k, z)$ (the remaining case is similar). In the first situation $(2[x-k], 2y, 2z)$ is a red position which we obtain by removing $2k-1$ sticks. In the second situation $(2x+1, 2[y-k], 2z)$ is a red position which we obtain by removing $2k-1$ sticks.

4. The only case that we must finally consider is a position of the form $(2x+1, 2y+1, 2z+1)$; this is always a black position since each of the entries is odd. How can we move to a red position?

If (x,y,z) is balanced then there is an obvious move: take away 1 from the first pile to produce the red position $(2x, 2y+1, 2z+1)$. If, however, (x,y,z) is unbalanced we can balance it to the position $(x-k, y, z)$ (the remaining cases are similar). Then $(2[x-k], 2y+1, 2z+1)$ is a red position which we obtain by removing $2k-1$ sticks.

Conclusion   Our analysis shows that we can win the game, starting from a black position, since we can always find a way to produce a red position and our opponenet must always produce a black position. Eventually we end up with the position $(0,0,0)$ which is a red position and we win. This is exactly the same as the balanced and unbalanced argument and shows that the red positions are simply the balanced positions and the black positions are the unbalanced one. So now we can drop the red and black language and go back to balanced and unbalanced.

We have not really solved the game, we have just found a convenient way of describing balanced positions in the language of near-doubling. A little more thinking about this, however, leads to an elegant solution.

3.5   Nim solved by near-doubling

We can now easily generate all the balanced position in a three-pile Nim game using the near-doubling argument. For example, starting with the balanced position $(0,0,0)$ we can construct all of its near-doubles

$$(1,1,0),\ (1,0,1),\ \text{and}\ (0,0,1)$$

and then all near-doubles of those three positions. All such positions must be also balanced. By continuing in this way we see that all of these positions are balanced:

$$(1,1,0),\ (1+2,2,1),(1+2+2^2,1+2^2,2),$$
$$(1+2+2^2+2^3,2+2^3,1+2^3),(1+2+2^2+2^3,1+2+2^3,2^3),\ldots$$

This is faster than starting at the bottom and directly computing balanced positions by our other method. But it is still somewhat strange-looking.

At some point in these investigations, now or perhaps a bit earlier, we must begin to see that our perception of the problem has been clouded by using the decimal arithmetic notation. Certainly this pattern demands a *binary interpretation*. These examples suggest it. Near-doubling suggests it.

An elegant strategy for Nim will become transparent provided we switch to a binary representation of the piles. For example, the position

$$(1,2,5,7,11),$$

written in decimal notation, is far less informative to us than when written in binary notation. Doubling or near-doubling this position in decimal notation is a tedious exercise in arithmetic that does not reveal much. Doubling or near-doubling this position in *binary* is surprisingly simple and revealing. The reader is invited to review binary arithmetic (covered now in Section 3.5.1) before returning to this in Example 3.5.3.

3.5.1 Review of binary arithmetic

We provide now a quick review of how numbers can be expressed in different bases. This section may be omitted by any reader who feels comfortable with base 2 arithmetic and is eager to apply it to the Nim game.

Suppose we have 147 eggs. What does the notation "147" really mean? One way of understanding the notation is as follows. If we put our eggs into boxes of ten eggs each, we would have fourteen boxes and seven eggs left over. These seven eggs account for the numeral "7".

We now put the fourteen boxes into crates which hold ten boxes each. We fill up one full crate and have four boxes left over. These four boxes account for the numeral "4". Since there are fewer than ten crates, we need not do any further grouping. We have one crate left over and this accounts for the numeral 1.

But egg boxes usually hold twelve eggs each. If a crate contains twelve boxes, we could easily check that we would have one crate, no loose boxes, and three loose eggs; this would represent the number of eggs in base twelve. Our process of arriving at the numeral 103 can be looked upon as successive division and recording remainders: if we divide 147 by 12 we get 12 with a remainder of 3, thus accounting for the "3" in "103". If we then divide 12 by 12, we get 1 with a remainder of zero, thus accounting for the "0" and the "1".

We can do the same computation relative to any positive integer $n \geq 2$, thus arriving at a base n numeral for the (base ten) number 147. We simply divide by n and record the remainder, then divide the partial quotient by n and record the remainder, etc. We continue the process until the final partial quotient is zero.

Example 3.5.1 We illustrate with $n = 2$ and we work again with the number 147 as our starting point.

$$
\begin{aligned}
147 \div 2 &= 73 && \text{with remainder } 1 \\
73 \div 2 &= 36 && \text{with remainder } 1 \\
36 \div 2 &= 18 && \text{with remainder } 0 \\
18 \div 2 &= 9 && \text{with remainder } 0 \\
9 \div 2 &= 4 && \text{with remainder } 1 \\
4 \div 2 &= 2 && \text{with remainder } 0 \\
2 \div 2 &= 1 && \text{with remainder } 0 \\
1 \div 2 &= 0 && \text{with remainder } 1
\end{aligned}
$$

This all works out to the notation 10010011 meaning

$$1 \cdot 2^7 + 0 \cdot 2^6 + 0 \cdot 2^5 + 1 \cdot 2^4 + 0 \cdot 2^3 + 0 \cdot 2^2 + 1 \cdot 2^1 + 1$$

where the final remainder is the left-most *bit* and the first remainder is the right-most *bit*.

Thus 147 (base 10) equals 10010011 (base 2). Note that this is really similar to the meaning of 147 (base 10):

$$1 \cdot 10^2 + 4 \cdot 10 + 7.$$

In fact we have verified an unusual looking statement, namely that

$$1 \cdot 10^2 + 4 \cdot 10 + 7 = 1 \cdot 2^7 + 0 \cdot 2^6 + 0 \cdot 2^5 + 1 \cdot 2^4 + 0 \cdot 2^3 + 0 \cdot 2^2 + 1 \cdot 2^1 + 1.$$

Both of these are just ways of writing the number we know as 147.

You may wish to check (just by ordinary arithmetic) that 147 can also be written as

$$147 = 1 \cdot 12^2 + 0 \cdot 12 + 3.$$

Thus our number one-hundred and forty-seven can be written as 147 (base 10), or as 10010011 (base 2) or even as 103 (base 12).

All of these are just different ways of writing the number which we usually call one-hundred and forty-seven (and the ancient Romans would have called CXLVII). ◀

Example 3.5.2 For practice, we do one more quick computation. To write the number twenty-six (i.e, 26 in the usual base 10) as a base 2 numeral, we observe that

$$
\begin{aligned}
26 \div 2 &= 13 \quad \text{with remainder } 0 \\
13 \div 2 &= 6 \quad\; \text{with remainder } 1 \\
6 \div 2 &= 3 \quad\; \text{with remainder } 0 \\
3 \div 2 &= 1 \quad\; \text{with remainder } 1 \\
1 \div 2 &= 0 \quad\; \text{with remainder } 1
\end{aligned}
$$

Thus

$$26 \text{ (base 10)} = 11010 \text{ (base 2)}.$$

Note that we do not need the numbers 13, 6, 3, and 1 except to continue the calculations. We need just the remainders 0, 1, 0, 1, and 1. *When read from bottom to top*, this is just the base 2 numeral for 26. While we are accustomed to reading numerals from left to right, the convenience of doing the computations in the way we did results in our reading these binary representations from bottom to top.

Writing in columns   We shall need frequently to write our binary numbers in *columns* rather than rows. Thus the number 26, which in standard binary notation becomes 11010, will be expressed as a column this way:

$$26 \text{ (base 10)} = 11010 \text{ (base 2)} = \begin{pmatrix} 1 \\ 1 \\ 0 \\ 1 \\ 0 \end{pmatrix}$$

Note that the column order is the exact reverse of the order in which we computed the bits in our computation above. We computed the bottom bit first and then all the other bits in order, from bottom to top.   ◄

3.5.2   Simple solution for the game of Nim

The near-doubling argument allows us to generate quickly and easily all the balanced positions in Nim. When we do this using binary notation the structure becomes almost obvious.

Example 3.5.3 A position in the game of Nim written in decimal notation as

$$(1,2,5,7,11)$$

appears in binary notation as

$$(1, 10, 101, 111, 1011)$$

or, if we prefer to arrange the bits in columns, we can display the position as

0	0	0	0	1
0	0	1	1	0
0	1	0	1	1
1	0	1	1	1

Near doubling is easy now. Multiplying by two simply raises the rows in the display:

0	0	0	0	1
0	0	1	1	0
0	1	0	1	1
1	0	1	1	1
0	0	0	0	0

To add a pair of 1's (or two pairs of 1's) add them to the bottom row.   ◄

Example 3.5.4 We already know that $(1,2,3)$ is a balanced position. We express this position in binary:

0	0	0
0	1	1
1	0	1

This position is a near-double of the position

0	0	0
0	0	0
0	1	1

which itself is a near-double of the end position

0	0	0
0	0	0
0	0	0

Note that there are an even number of 1 bits in each row. Doubling merely adds a row of zeros to the bottom. Near-doubling does the same, but adds possibly two 1 bits to the bottom row. Even if we do this thousands of times, one thing is transparent: there will always be an even number of 1 bits in every row.   ◀

In fact we can prove this fact in complete generality.

3.5.5 *A position in 3-pile Nim is balanced if and only if there are an even number of 1 bits in each row.*

All balanced positions can be generated by starting with the balanced position $(0,0,0)$ and doing near-doubling repeatedly. Every near-double has an even number of 1 bits in each row. That is all there is to the proof.

This also gives us our game strategy. If a position is unbalanced it is because there is an odd number of bits in one or more rows. At least one of the columns will allow a reduction of sticks in its associated pile so as to produce an even number of bits in each row. Binary arithmetic shows how.

3.5.3   Déjà vu?

Haven't we seen this before? The game of binary bits in Section 3.3 looks identical to this. In the game of binary bits the balanced positions were exactly the same: even number of 1 bits in each row. Is it possible? Why didn't we notice this before?

3.5.6 *The game of Nim is equivalent to the game of binary bits.*

We need to check that the rules of Nim and the rules of binary bits are the same. Certainly the positions are the same.

In the game of Nim $(n_1, n_2, \ldots, n_k)$ the rules require us to select a pile and reduce the pile by one or more sticks. If we convert each of the numbers to binary and use them to play a game of binary bits the rules require us to select a binary bit 1 in some column to change to 0 and then change (as we please) all the bits below it. If we remember how binary arithmetic works we see that this is equivalent to reducing the Nim number that corresponds to that column. The Nim game ends with a position of $(0,0,\ldots,0)$ while the binary bits game ends with no 1 bits just 0 bits. The last player to move wins.

The two games are identical. Thus, since we have an easy solution of the binary bits game, we have an obvious strategy for Nim: *convert every Nim game to a binary bits game.*

Example 3.5.7 The position $(1,2,5,7,11)$ in 5-pile Nim is unbalanced. (Not so easy to see.) The balancing move is to take 10 sticks from the last pile. (Really not at all easy to see.) There is only one balancing move. (Why?)

The answer to our difficulties is to play binary bits instead where everything is truly easy to see. Convert $(1,2,5,7,11)$ to $(1,10,101,111,1011)$ in binary. Now display this position as in Figure 3.15. Here we have entered each of the

$$\begin{array}{ccccc} 0 & 0 & 0 & 0 & 1 \\ 0 & 0 & 1 & 1 & 0 \\ 0 & 1 & 0 & 1 & 1 \\ 1 & 0 & 1 & 1 & 1 \end{array}$$

Figure 3.15: The position $(1,2,5,7,11)$ displayed in binary.

binary expressions for the numbers 1, 2, 5, 7, and 11 as binary columns. Taking sticks from any one of these five piles is the same as a legal binary bits move on one of these five columns.

Figure 3.16 shows the correct balancing move in this game of binary bits. This corresponds to the move

$$(1,2,5,7,11) \rightsquigarrow (1,2,5,7,1)$$

in our 5–pile Nim game.

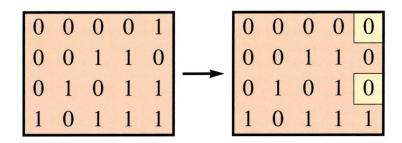

Figure 3.16: The move $(1,2,5,7,11) \rightsquigarrow (1,2,5,7,1)$ displayed in binary.

This makes it clear why the balancing move is to take 10 sticks from the last pile. We can also see at once that this is the only balancing move possible. A 5–pile game would have defeated us before. Now the play in binary notation is straightforward (depending on your skills with binary arithmetic).   ◀

Problem 128 *Under what conditions will it be possible to find more than one balancing move in a three–pile Nim game? How many balancing moves will there then be?*   Answer □

Problem 129 *Find all three ways in which the game (9,11,13) can be balanced.*
Answer □

Problem 130 *Can there be more than 3 balancing moves in a 3-pile Nim game?*
Answer □

Problem 131 *In a 10–pile unbalanced Nim game, what is the largest possible number of balancing moves?*
Answer □

Problem 132 *In a 11–pile unbalanced Nim game, what is the largest possible number of balancing moves?*
Answer □

Problem 133 *How many different balancing moves are there for the Nim game $(1,3,5,7,9,11,13,100000)$?*
Answer □

Problem 134 *How can you tell immediately that the Nim game $(136,72,48,40)$ is unbalanced? Can you spot the pile that needs adjusting without much computation?*
Answer □

Problem 135 (Opening strategy) *You are invited to a game of Nim and you spot that the opening position is balanced. Your opponent invites you to start. What do you do?*
Answer □

Problem 136 (Poker Nim) *You are invited to a game of Nim played with coins and with a new rule. The coins are placed in three piles and each player, in turn, may take as many coins as he likes from a single pile and keep them aside. At any play of the game a player may decide to return coins from his collection as he wishes, and place them on a pile, instead of reducing a pile. The player who takes the last coin wins and seizes all the coins on the table. Discuss.*
Answer □

3.6   Return to marker games

Let us return now to the marker games. We have already established that the 2, 3 and 4–marker games have balanced positions which can easily be described in terms of simple equations:

$$B = A + 1$$

for the 2–marker game,

$$A + B = C$$

for the 3–marker game, and

$$D - C = B - A$$

for the 4–marker game. But we did not see what to do for the 5 or 6–marker games. It appears to be obvious that we should search harder, much harder, to find the correct equations for the n–marker games!

But that would be misguided. There is a pattern which generalizes from 2, 3 and 4–marker games to all n–marker games but this is not it. Many times in mathematics the attempt to generalize something requires a new way of looking at the simpler cases.

Let us pause and reflect on this for a moment. The strategy for 2–pile Nim is the mirror strategy (the Tweedledee-Tweedledum strategy). Had we insisted on finding some kind of mirror strategy for 3–pile Nim we would have surely failed. Instead we came up with the strategic device of converting the numbers to binary. That allowed us to solve all Nim games. Had we looked at the 2-pile Nim game in binary we might have noticed that the mirror strategy was really all about ensuring an even number of 1 bits in each row. We missed the chance to find the pattern that works for all Nim games because we were looking too hard at the wrong pattern.

The same is true for marker games. By looking too closely at the balancing equations for 2, 3 and 4–marker games we completely miss the perspective that will allow us to solve all Marker games.

3.6.1 Mind the gap

The perspective that we need for marker games involves viewing a move as a *closing* or *opening* of a gap. A *gap* is just the number of holes—legal positions— between a pair of markers. Once we have expressed the objective of the game in the language of gaps and openings and closings, rather than equations, we will find a pattern that works in general.

Example 3.6.1 [Gaps in a 3–marker game] Place markers A, B and C at 4, 9 and 13 as in Figure 3.17.

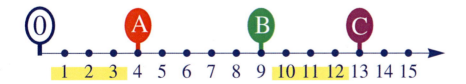

Figure 3.17: Gaps in the 3–marker game with markers at 4, 9, and 13.

We know that the position is balanced if $A + B = C$, which we can rewrite as

$$A - 0 = C - B.$$

To interpret this in the language of gaps we would prefer to write

$$A - 1 = C - B - 1.$$

There are gaps between 0 and *A*, between *A* and *B* and between *B* and *C*. Of these it is only the first and third gap that concern us. Both gaps are equal to 3 as we see by counting (or, equivalently, by computing $A - 0 - 1$ and $C - B - 1$).

Thus we have a balanced position corresponding to the position $(3,3)$ in Nim. If our opponent moves *A*, this reduces a gap and we answer by moving *C* the same number of places. If the opponent moves *B*, this widens a gap, and we answer by moving *C* the same number of places. If our opponent moves *C*, this reduces a gap and we answer by moving *A* the same number of places.   ◀

Example 3.6.2 [Gaps in a 4–marker game] Place markers *A*, *B*, *C*, and *D* at 5, 10, 20, and 30 (as in Figure 3.18). The balancing move would be to move *D* from 30 to 25, for in that case we would have

$$B - A = D - C = 5,$$

a balanced position. Expressing this in terms of the two gaps we would have

$$B - A - 1 = D - C - 1 = 4,$$

so that the number of holes between *A* and *B* is the same as the number of holes between *D* and *C*. We have closed the gap between *C* and *D* to the same size as the gap between *A* and *B*. Note that it is only these two gaps that matter; the other gaps (between 0 and *A* or between *B* and *C*) do not interest us.

Figure 3.18: Gaps in the 4–marker game with markers at 5, 10, 20, and 30.

The balancing equation is really demanding that we develop a mirror strategy (Tweedledee-Tweedledum strategy) that focuses instead on the two gaps. That means, too, that there is a similarity between the strategies for the 4–marker game and the 2–pile Nim game. The games themselves are not identical. The point is that a position in a 4–marker game can be balanced by comparing it with a related position in a 2–pile Nim game.   ◀

Problem 137 *Explain the similarity between the strategies for the 4–marker game and the 2–pile Nim game?*   Answer □

Problem 138 *Formulate a similar analogy between the 3-marker game and the 2–pile Nim game.*   Answer □

Problem 139 *Formulate a similar analogy between the 2-marker game and the 1–pile Nim game.*   Answer □

Problem 140 *Use the strategy for 3–pile Nim to find strategies for the 5 and 6–marker games. Note that these strategies do not involve simple equations similar to those which arose in the lower order marker games.*   Answer □

3.6.2   Strategy for the 6–marker game

Once it occurs to us that we can use the gaps to compare our position to a Nim game the solution is simple. We do not even have to do much more thinking about it.

Let us consider first the 6-marker game. Designate the markers, from left to right, by A, B, C, D, E, and F. The gaps (number of empty holes) between A and B, between C and D and between E and F give us three numbers, x, y and z, which correspond to a certain Nim game (x, y, z) as indicated in Figure 3.19.

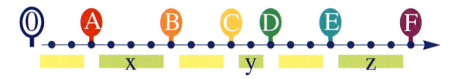

Figure 3.19: The three key gaps in the 6–marker game.

Since we are counting holes in between,

$$x = B - A - 1,$$
$$y = D - C - 1,$$

and

$$z = F - E - 1.$$

The argument   Define a position in the 6–marker game to be a *red position* if the gaps (x, y, z) correspond to a balanced position in 3–pile Nim. Call the position *black* if this is not so. Then

1. The final position in the 6–marker game is a red position. This is because $(0, 0, 0)$ is a balanced position in Nim.

2. Any move from a red position will result in a black position. This is because any such move will change one of the markers and so change exactly one of the gaps. So if (x, y, z) is balanced in Nim, the new set of gaps must be unbalanced in Nim.

3. Given a black position there is a move of markers that produces a red position. A black position corresponds to a gap triple (x, y, z) that is unbalanced in Nim. Find a balancing move in Nim and then move the appropriate marker to produce a new balanced set of gaps.

The final position is red, red always moves to a black, and from a black one can find at least one move to a red. It follows that the red positions are the balanced positions in the 6–marker game and the black positions are all unbalanced.

A small subtle point   This argument shows that we have captured all of the balanced positions by comparing to Nim. But it does not say that the two games are identical.

If a position corresponds to the gaps (x, y, z) and one of the markers B, D, or F is moved, then indeed the new gap position does correspond to a move in Nim because one of the numbers x, y or z has been reduced. But if one of the markers A, C, or E is moved the effect is that of *widening* a gap. This does not correspond to a move in the associated Nim game. In Nim we change numbers only by reducing them.

But this doesn't impede our play. We just move the marker at the other end of the gap to restore its previous size. This leads to the same balanced Nim game that existed before our opponent made his move.

Example 3.6.3 Consider the marker game with markers at

$$5, 7, 12, 15, 20, \text{ and } 24$$

as in Figure 3.20.

Figure 3.20: The 6–marker game with markers at 5, 7, 12, 15, 20, and 24.

The gaps are of sizes 1, 2 and 3 respectively. Thus, we look at the associated Nim game $(1, 2, 3)$. We remember that this as a balanced Nim position. (If we do not, we could write the numbers out in binary and check.)

Our opponent makes a move: say, he or she moves the marker E from 20 down to 18. This widens a gap, so does *not* correspond to a Nim move. Even so, Nim helps us rebalance.

Our answer is to move marker F from 24 to 22. The markers are now at 5, 7, 12, 15, 18, and 22. This position once again corresponds to the Nim game $(1, 2, 3)$. Now, suppose, our opponent moves marker D from 15 down to 13. This reduces a gap, so it *does* correspond to a Nim move: the markers are now at 5, 7, 12, 13, 18 and 22, and this corresponds to the Nim position $(1, 0, 3)$. This Nim position is unbalanced and we could balance it by taking 2 sticks from the third pile, leaving the balanced Nim position $(1, 0, 1)$.

This Nim move would correspond to the move in the marker game in which we move marker F from 22 to 20. The markers are now at 5, 7, 12, 13, 18, and 20. This is a balanced position because it corresponds to the Nim game $(1, 0, 1)$,

◄

3.6.3 Strategy for the 5–marker game

For the five marker game, the analysis is the same in all details except that our gaps are determined by the number of holes to the left of A, the number of holes between B and C and the number of holes between D and E give us three numbers, x, y and z, which correspond to a certain Nim game (x,y,z) as indicated in the sketch:.

$$0 \leftarrow x \rightarrow A \longleftrightarrow B \leftarrow y \rightarrow C \longleftrightarrow D \leftarrow z \rightarrow E.$$

Since we are counting holes in between,

$$x = A - 1, \; y = C - B - 1, \; \text{and} \; z = E - D - 1.$$

Example 3.6.4 Consider the marker game with markers at

$$5, \; 10, \; 14, \; 20 \text{ and } 22.$$

as in Figure 3.21.

Figure 3.21: The 5–marker game with markers at $5, \; 10, \; 14, \; 20$ and 22.

This corresponds to the Nim game $(4,3,1)$. This Nim position is unbalanced and could be balanced by taking 2 sticks from the first pile, leaving the position $(2,3,1)$. This would correspond to moving marker A from 5 to 3, leaving the markers at 3, 10, 14, 20, and 22. ◄

3.6.4 Strategy for all marker games

With more than 6 markers, the analyses are similar. A marker game with an even number of markers, say $2n$, corresponds to a Nim game of n piles. A marker game with an odd number of markers, say $2n - 1$, also corresponds to a Nim game of n piles. One must only remember that the number of holes between the successive pairs of markers determines the associated Nim game, and that if the number of markers is odd, our first gap is that between 0 and A.

Problem 141 *The marker game with markers at*

$$10, \; 15, \; 20, \; 25, \; 40, \; 50, \; 60 \text{ and } 80$$

as in Figure 3.22 corresponds to what Nim game?

Answer □

Figure 3.22: An 8–marker game.

Problem 142 *The marker game with markers at*

$$10,\ 15,\ 20,\ 25,\ 40,\ 50\ and\ 60$$

corresponds to what Nim game?   Answer ☐

Problem 143 *Find all balancing moves in the game with markers at 5, 9, 13, 14, 20 and 27.*   Answer ☐

Problem 144 *Find all balancing moves in the game with markers at 5, 9, 13, 14 and 20.*   Answer ☐

Problem 145 *Find all balancing moves in the game with markers at 5, 9, 13, 14, 20, 27, 33 and 100.*   Answer ☐

Problem 146 *Find all balancing moves in the game with markers at 5, 9, 13, 14, 20, 27, 33, 100 and 200.*   Answer ☐

Problem 147 *Is it possible that an 8–marker unbalanced game could have more than 4 balancing moves? Explain.*   Answer ☐

3.7 Misère Nim

In our two-player game of Nim the player who takes the last stick *wins*. In the *Misère* version of Nim, the player who is forced to take the last stick *loses*. It is the Misère version of the Nim game that plays a mysterious and recurring role in Alain Renais's cult 1961 film *Last Year at Marienbad* where the piles take the form of rows of cards.

Figure 3.23: *Last Year at Marienbad.*

Problem 148 *Find a winning strategy for the game of Misère Nim.*   Answer □

3.8   Reverse Nim

A student in one of our classes suggested a variant of Nim. In this variant there are several piles of match sticks. The two players move alternately and the one who takes the last stick wins. The difference is that in this game, a player may take as many sticks as he or she wishes, but at most one from each pile. Thus, when it is your move you may take a single stick from each of as many piles as you like but you must take at least one stick.

There is also a *misère* version of this game. The rules for Reverse Misère Nim are the same except that the one who takes the last stick *loses*.

Problem 149 *Find a strategy for this game of Reverse Nim.*   Answer □

Problem 150 *Find a strategy for this game of Reverse Misère Nim.*

Answer □

3.8.1   How to reverse Nim

We already know a simple strategy for Reverse Nim (Problem 149) but if we revisit this problem it will help in finding a strategy for the misère version of the game. We illustrate by considering a Reverse Nim game $(7,5,3,1)$ as in Figure 3.24.

Arrange the piles in a more suggestive format as in Figure 3.25. Here there are two perspectives on the same position as the shading suggests. In the first perspective we see rows and in the second we see columns.

Figure 3.24: A Reverse Nim game with 4 piles.

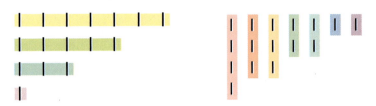

Figure 3.25: Two perspectives on Reverse Nim game with 4 piles.

Let us take the viewpoint that each *column* is a pile. We now have seven piles of sticks. The rules of Reverse Nim translate into allowing us to take as many sticks as we like as long as we take them all from the same vertical pile. (We must take at least one stick, of course.) So our Reverse Nim game $(7,5,3,1)$ translates to the ordinary 7–pile Nim game

$$(4,3,3,2,2,1,1).$$

Figure 3.26 shows this position along with the necessary computation in binary that allows us to recognize the position as unbalanced in 7–pile Nim.

(4,	3,	3,	2,	2,	1,	1)
1	0	0	0	0	0	0
0	1	1	1	1	0	0
0	1	1	0	0	1	1

Figure 3.26: Playing the associated 7–pile Nim game.

There is one balancing move (in 7–pile Nim), namely to take all four sticks in the first (column) pile leaving

$$(0,3,3,2,2,1,1).$$

The new display of sticks (in Reverse Nim) is

$$(6, 4, 2)$$

since we have taken one stick from each of the four piles of the original Reverse Nim problem. This is illustrated in Figure 3.27

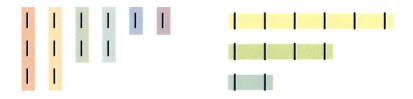

Figure 3.27: After the balancing move.

We can continue in this way, going back and forth between the Reverse Nim position and the corresponding Nim position, making our move in Nim and interpreting it in reverse Nim. Of course, we didn't need to go to all this trouble to achieve a balanced position because we had already observed (in Problem 149) that the balanced positions are those in which all piles have an even number of sticks, so it was obvious that in the original Reverse Nim problem all we had to do was take one stick from each pile. But our perspective allows us to understand the structure of the game in a way that could be useful for determining a strategy for the Reverse Misère version of the game. What is this strategy?

3.8.2 How to play Reverse Misère Nim

While playing Reverse *Misère* Nim we simply shift our perspective as we just did for Reverse Nim. Consider each column as a pile and use the strategy that we developed for Misère Nim in Section 3.7 on the resulting Misère Nim game, repeatedly translating our results back and forth between the Reverse Misère Nim positions and the corresponding Misère Nim positions.

Observe this offers another example of a case in which a simple solution to a problem does not reveal enough to obtain solutions to closely related problems. Here the easy solution to Reverse Nim offered little help in solving Reverse Misère Nim.

We saw an additional example with our marker games. The simple solutions involving algebraic equations to identify the balanced positions for such games with four or fewer markers may have suggested similar equations for games involving more than four markers. But that led us in the wrong direction. Once we understood Nim, however, it was easy to make the correct connection to the marker games.

We see similar situations in many parts of mathematics. For example, in our chapter on Links, an easy solution to constructing certain configurations will not point the way to obtaining constructions of configurations that are slightly more complex. We will need another perspective for that. The key concept involves introducing a new idea. Once the new idea has been formulated, the method of proceeding becomes clear.

3.9   Summary and Perspectives

We obtained complete strategies for two rather complicated games: Nim and the Marker Games. Several aspects of creative mathematics and discovery appeared in our developments.

1. We started with very simple versions of the games (the low marker games and 2–pile Nim). This gave us a *feel* for the game and helped us "discover" the concept of *balanced positions* which was central to all of our games.

2. Our experience with some easy 3—pile Nim games was useful in obtaining strategies for more complicated 3—pile games. The big step was our recognizing that the strategies had something to do with *doubling* or *near-doubling* of piles and eventually we made the connection with base 2 arithmetic and recognized Nim as equivalent to the binary bits game.

3. Once we understood 3–pile Nim, it was a small step to understand Nim with any number of piles. But, our understanding of the 2, 3 and 4 marker games appeared to offer no help towards understanding marker games with more than 4 markers. This was so because we focused on the wrong thing: the position of the markers instead of the size of the gaps. Surprisingly, the marker game turned out to be closely related to the game of Nim—the relationship was so close, in fact, that our strategy for Nim allowed us to determine a strategy for the marker games.

In these observations is an example of something that often occurs in mathematics. Two seemingly unrelated problems lend themselves to similar mathematical analysis. What this means, of course, is that the two problems really have a similar underlying structure, even though the two problems may superficially appear to be unrelated. Another example is a surprising connection that exists between our material on tiling in Chapter 1 and electricity.

Some of the material in this section, and a very nice treatment of various other games, can be found in the book *Excursion into Mathematics* by Beck, Bleicher and Crowe (item [1] in our bibliography).

3.10   Supplementary material

We conclude our chapter with some supplementary material that the reader may find of interest in connection with our study of the game of Nim.

3.10.1 Another analysis of the game of Nim

Our analysis of the game of Nim is close to the original ideas of Bouton when he solved the game in 1902. The game was revisited in the 1930's by R. P. Sprague and P. M. Grundy independently. Mathematicians often revisit old problems trying to find new perspectives and possible generalizations. Some call the process *squeezing the lemon*. If you have ever squeezed a lemon you well know that you can always find at least one more drop.

Let us go back to one-pile Nim and two-pile Nim. These games were very easy to solve but it is not the *solution* that we want to revisit, but the *nature* of the games. Curiously, two-pile Nim looks to be just two games of one-pile Nim. You could describe the rules as requiring each player at his turn to select one of the one-pile games and play a move in that. The game ends when there are no moves to be made in either of the one-pile games. Two-pile Nim is just the sum of two games of one-pile Nim.

Do notice, however, that adding two games produces some interesting complexities. The strategy for one-pile Nim does not help at all in determining the strategy for two games of one-pile Nim added together.

Adding two games Suppose that we have two games G_1 and G_2 of our familiar type: in each game players take turns moving and the winner is declared by the player who made the last legal move. We can produce a new game $G_1 + G_2$ called the *sum of the two games* by making this rule: each player at his turn is to select one of the two games and play a legal move in that game. The game ends when there are no moves to be made in either of the two games and the player to make the last move wins.

For example a two-pile Nim game $(7, 10)$ would be the sum of the one-pile Nim game (7) and the one-pile Nim game (10). Similarly the classic Nim game $(1, 3, 5, 7)$ is the sum of the four one-pile Nim games (1), (3), (5), and (7). Or, if you prefer, it is the sum of the two two-pile Nim games $(1, 3)$ and $(5, 7)$.

Our goal in studying this game summing idea is to find out how information about the separate games G_1 and G_2 can be used to find a strategy for the game $G_1 + G_2$.

3.10.2 Grundy number

The first element of wishful thinking we can dispense with easily. Even if we know all the balanced and unbalanced positions for the games G_1 and G_2 this in no way helps us find the balanced and unbalanced positions for the game $G_1 + G_2$. We saw this in our study of Nim. Even though a Nim game may be thought of as a sum of smaller-pile Nim games, we found that solving one-pile Nim did not help solve two-pile Nim, nor did solving both of these help in solving three-pile Nim.

In Section 3.2.6 we described balanced positions in a game using the computation of the balancing number Balance(p). That balancing number just computes as 0 or 1 depending on whether the position p in the game is balanced or unbalanced. The computation loses all other information. The clever idea of Sprague and Grundy was to adjust this to reflect just how far a position might be "distant" from a balanced position.

The definition   Here is the definition. Note how closely it follows the way we defined the balancing number Balance(p) for positions in a game in Section 3.2.6.

1. If p is an end position in the game G then Grundy$(p) = 0$.

2. If p is not an end position in the game G then find all the positions
$$p \rightsquigarrow p_1, p_2, \ldots, p_n$$
that follow from one legal move, and compute the list of numbers
$$\text{Grundy}(p_1), \text{Grundy}(p_2), \ldots, \text{Grundy}(p_n).$$
Then Grundy(p) is defined to be the smallest of the numbers $0, 1, 2, 3, \ldots$ that does *not appear* in this list.

Thus the Sprague-Grundy function Grundy(p) assigns a number to each position p in the game. We start at the bottom. Any ending position has a value of zero. So, if p_0 is an end position, then Grundy$(p_0) = 0$. For any other position p in the game we look for all possible positions q that can follow p by a single move. Then Grundy(p) is defined to be the smallest integer that is not the same as one of the values Grundy(q) for some position q that can follow p. Thus write out
$$0, 1, 2, 3, 4, 5, 6, \ldots$$
and strike out the ones that have appeared for a position following after p. Take the smallest that is left. You have to take Grundy$(p) = 0$ if none of the next positions has a zero value. (This is why balanced positions will have Grundy$(p) = 0$.)

This is an example of a recursive definition; we have to build up the values of the function Grundy(p) step by step starting close to end of the game.

Note that
$$\text{Grundy}(p) = 0 \text{ if and only if } \text{Balance}(p) = 0$$
and
$$\text{Grundy}(p) \geq 1 \text{ if and only if } \text{Balance}(p) = 1.$$

Problem 151 *Compute the values of the Sprague-Grundy function for a position in a one-pile game of Nim.*

Answer □

Problem 152 *Let us play the Nim game* $(1,2,3)$. *Compute the Sprague-Grundy function for all the positions*

$$(0,0,0),\ (0,0,1),\ (0,1,0),\ (1,0,0), (0,1,1),\ \ldots, (1,2,2),\ (1,2,3)$$

in the game.   Answer □

Problem 153 *See if you can discover the exact formula for the Sprague-Grundy function for a position in a two-pile game of Nim. Write*

$$\mathrm{Grundy}(m,n) = m \oplus n$$

and find what this operation must be. This is called the Nim-sum and is explained in detail in Section 3.10.3 below. You may succeed in spotting how to compute this. [Hint: Look at the numbers in binary.]   Answer □

Problem 154 *If you succeeded in determing how the operation* $m \oplus n$ *works then give a try at proving that the Grundy number for a position* (m,n) *in 2–pile Nim is exactly the Nim-sum* $m \oplus n$. *Use induction on the depth of the position.*
                                                                     Answer □

3.10.3   Nim-sums computed

Binary addition *without carry* is a special case of bitwise addition where

$$0 + 0 = 0$$
$$1 + 0 = 1$$

and

$$1 + 1 = 0.$$

That leads to the notion of a nim-sum. We define the sum $m \oplus n$ to be be the number obtained by summing m and n (expressed in binary) but adding the binary bits without carry.

Example 3.10.1 Let us perform the computation

$$7 \oplus 5 = 2.$$

In decimal it looks rather mysterious. If we write in binary instead

$$111 \oplus 101 = 10$$

the pattern is clearer. Not clear enough? How about

$$\begin{bmatrix} 1 \\ 1 \\ 1 \end{bmatrix} \oplus \begin{bmatrix} 1 \\ 0 \\ 1 \end{bmatrix} = \begin{bmatrix} 0 \\ 1 \\ 0 \end{bmatrix} ?$$

◀

Problem 155 *Do some of the computations in Figure 3.28.*   □

⊕	1	2	3	4	5	6	7	8	9	10
1	0	3	2	5	4	7	6	9	8	11
2	3	0	1	6	7	4	5	10	11	8
3	2	1	0	7	6	5	4	11	10	9
4	5	6	7	0	1	2	3	12	13	14
5	4	7	6	1	0	3	2	13	12	15
6	7	4	5	2	3	0	1	14	15	12
7	6	5	4	3	2	1	0	15	14	13
8	9	10	11	12	13	14	15	0	1	2
9	8	11	10	13	12	15	14	1	0	3
10	11	8	9	14	15	12	13	2	3	0

Figure 3.28: An addition table for ⊕.

3.10.4 Proof of the Sprague-Grundy theorem

Readers with more mathematical background could benefit from reading the proof of the Sprague-Grundy theorem. It is considerably longer than any of the previous proofs in this chapter, but it illustrates how a proof in a more advanced book might look. It reveals the structure of some of the types of games we studied in this chapter.

We have already studied a special case of this. In Problem 153 we discovered that the 2-pile Nim game (m, n), which is the sum of the two one-pile games (m) and (n), has the Grundy value equal to $m \oplus n$.

3.10.2 (Sprague-Grundy theorem) *The Grundy numbers for the sum of two games can be written in the form*

$$\mathrm{Grundy}(p_1, p_2) = \mathrm{Grundy}(p_1) \oplus \mathrm{Grundy}(p_2).$$

where $\oplus$ is the nim-sum operation.

Depth of a game How far are we from the bottom of the game? A game with no moves has depth 0. A game where all moves lead immediately to the end position has depth 1. In this way we can define depth for any position in the game. (Depth of a position is defined in Section 3.2.6, but it is enough to see this intuitively for our proof.) This allows us to use induction on the depth of a game. Usually the statement we want to prove is obvious at depth zero, so the induction starts off easily.

Proof of the theorem At depth zero the theorem is evidently true, since it amounts only to the fact that $0 \oplus 0 = 0$. Thus it is only the induction step that takes us some trouble. Our proof below uses the assumption that we already know the theorem is true at any lower depth.

Let
$$b = \text{Grundy}(p_1) \oplus \text{Grundy}(p_2)$$
In order for us to prove that $\text{Grundy}(p_1, p_2) = b$ we must show that both of these statements are true:

1. For every non-negative integer $a < b$, there is a follower of (p_1, p_2) in the sum game that has Grundy value a.

2. No follower of (p_1, p_2) has the Grundy value b.

Then the Grundy value at (p_1, p_2), being the smallest value not assumed by one of its followers, must be b.

To show (1), let $d = a \oplus b$ and let k be the number of digits in the binary expansion of d, so that
$$2^{k-1} \leq d < 2^k$$
and d has a 1 bit in the kth position in the binary expansion.

We have to remember now that $d = a \oplus b$ and remember too how the binary without carry operation $\oplus$ works. Since $a < b$, b must have a 1 in the kth position and a must have a 0 there. Since
$$b = \text{Grundy}(p_1) \oplus \text{Grundy}(p_2)$$
we see that p_1 [or perhaps p_2] would have to have the property that the binary expansion of $\text{Grundy}(p_1)$ [or perhaps $\text{Grundy}(p_2)$] has a 1 in the kth position.

Suppose for simplicity that it is the first case. Then
$$d \oplus \text{Grundy}(p_1) < \text{Grundy}(p_1).$$
Now we have to remember what it means for a number to be smaller than a Grundy number. We would know that there is a move from p_1 to a position p_1' with that smaller number as its Grundy number, i.e., that
$$\text{Grundy}(p_1') = d \oplus \text{Grundy}(p_1).$$
Then the move from (p_1, p_2) to (p_1', p_2) is a legal move in the sum game and
$$\text{Grundy}(p_1') \oplus \text{Grundy}(p_2) = d \oplus \text{Grundy}(p_1) \oplus \text{Grundy}(p_2) = d \oplus b = a.$$
We have produced the move
$$(p_1, p_2) \rightsquigarrow (p_1', p_2)$$
for which
$$\text{Grundy}(p_1') \oplus \text{Grundy}(p_2) = a.$$
Since this position is at a lower depth we know (by our induction hypothesis) that
$$\text{Grundy}(p_1', p_2) = \text{Grundy}(p_1') \oplus \text{Grundy}(p_2) = a.$$
Thus the follower (p_1', p_2) in the sum game has a Grundy number a. This verifies our first statement.

Finally, to show (2), suppose to the contrary that (p_1, p_2) has a follower with the same Grundy value. We can suppose that this involves a move in the first game. (The argument would be similar if it involved a move in the second game.)

That is, we suppose that (p_1', p_2) is a follower of (p_1, p_2) and that

$$\text{Grundy}(p_1', p_2) = \text{Grundy}(p_1') \oplus \text{Grundy}(p_2) = \text{Grundy}(p_1) \oplus \text{Grundy}(p_2).$$

(Here we have again used our induction hypothesis since the position (p_1', p_2) is at a lower depth.) Just like in ordinary arithmetic (using $+$ instead of as here $\oplus$) we can cancel the two identical terms and conclude that

$$\text{Grundy}(p_1') = \text{Grundy}(p_1).$$

But this is impossible since

$$p_1 \rightsquigarrow p_1'$$

in the first game and no position can have a follower of the same Grundy value.

That completes the proof at the induction step and so the theorem follows.

3.10.5   Why does binary arithmetic keep coming up?

To explain the nim-sum requires an analysis using binary arithmetic. Why does this binary beast come out every time we address some problem about Nim and rest of the games that we have studied? There is an explanation that we can sketch here.

First of all there is an algebraic structure that we may not have noticed. If $\mathcal{N}$ is the null game (i.e., the game with no legal moves) then it must have a Grundy number of 0. But for any other game G the sum game $G + \mathcal{N}$ and the sum game $\mathcal{N} + G$ are just the original game G. (The only legal moves in the sum game are the moves in G itself.) Consequently

$$0 \oplus n = 0 \oplus n = 0$$

for any integer n.

The second element of algebraic structure is that the games $G_1 + G_2$ and $G_2 + G_1$ are identical. Consequently

$$m \oplus n = m \oplus n$$

for any integers m and n.

The third element of algebraic structure is that the games

$$(G_1 + G_2) + G_3$$

and

$$G_1 + (G_2 + G_3)$$

are identical. Consequently

$$(m \oplus n) \oplus p = m \oplus (n \oplus p)$$

for any integers m, n, and p.

The final element of algebraic structure is that any position p in a game G gives rise to a balanced position (p, p) in the sum game $G + G$. This is because we can always win from a position (p, p) by playing the mirror strategy, Consequently

$$n \oplus n = 0$$

for any integer n.

That is a lot of algebraic structure. The words normally used to describe this structure (some of them familiar) are commutative, associative group with every element its own inverse. (We will see groups structures again elsewhere in this text.) If we describe this structure to an algebraist we will be told instantly that the group operation is simply 1-bit binary addition without carry.

3.10.6 Another solution to Nim

We have solved Nim by converting it to a binary bits game. We can also solve Nim by using Nim-sums.

3.10.3 (Sprague-Grundy solves Nim) *A position* $(n_1, n_2, n_3, \ldots, n_k)$ *in a k-pile Nim game is balanced if and only if*

$$n_1 \oplus n_2 \oplus n_3 \oplus \cdots \oplus n_k = 0.$$

This follows directly from the Sprague-Grundy theorem since the Grundy number for that position computed directly from the sum of k one pile games

$$(n_1), (n_2), \ldots, (n_k)$$

is

$$n_1 \oplus n_2 \oplus n_3 \oplus \cdots \oplus n_k.$$

Problem 156 *Use the Sprague-Grundy theorem to show that the Nim position* (m, n) *is balanced if and only if* $m = n$. Answer ☐

Problem 157 *Use the Sprague-Grundy theorem and Table 3.28 to show that the Nim position* $(1, 2, 3)$ *is balanced and* $(2, 3, 4)$ *is not.* Answer ☐

Problem 158 *Use the Sprague-Grundy theorem and Table 3.28 to find a balancing move for* $(2, 3, 4)$.

Answer ☐

3.10.7 Playing the Nim game with nim-sums

The easiest way to play the correct strategy in Nim is to convert all piles to binary and then play the game of binary bits. The other rather elegant way of playing the game is to use the nim-sum operation as the key. A position $(n_1, n_2, n_3, \ldots, n_k)$ in the game of Nim is balanced if and only if the nim-sum

$$n_1 \oplus n_2 \oplus n_3 \oplus \cdots \oplus n_k = 0.$$

The nim-sum operation then helps in computing the correct move to make in the game. Figure 3.28 on page 140 is useful in giving us the addition table that we would need to use (or memorize) if we wish to be skillful players.

We illustrate with a simple example. But be sure to try Problem 163 and Problem 164 to make sure you see a possible subtlety in the method.

Example 3.10.4 The game $(8, 10, 12)$ is unbalanced. What are all the balancing moves? We compute

$$8 \oplus 10 \oplus 12 = (8 \oplus 10) \oplus 12 = 2 \oplus 12 = 14.$$

We note that

$$8 \oplus 10 \oplus 12 \oplus 14 = 14 \oplus 14 = 0.$$

Thus the only possible moves in the game that will produce a balanced position are

$$(8, 10, 12) \to (8 \oplus 14, 10, 12) = (6, 10, 12),$$
$$(8, 10, 12) \to (8, 10 \oplus 14, 12) = (8, 4, 12),$$

and

$$(8, 10, 12) \to (8, 10, 12 \oplus 14) = (8, 10, 2)$$

All of these are legal Nim moves.   ◀

Example 3.10.5 Here is the same example but with the arithmetic argued in a different way. The game $(8, 10, 12)$ is unbalanced. What are all the balancing moves? We note that

$$(8 \oplus 10) \oplus 8 \oplus 10 = 0$$

and so we move

$$(8, 10, 12) \to (8, 10, [8 \oplus 10]) = (8, 10, 2).$$

Similarly

$$8 \oplus (8 \oplus 12) \oplus 12 = 0$$

and so we move

$$(8, 10, 12) \to (8, [8 \oplus 12], 12) = (8, 4, 12).$$

And finally

$$(10 \oplus 12) \oplus 10 \oplus 12 = 0$$

and so we move

$$(8, 10, 12) \to ([10 \oplus 12], 10, 12) = (6, 10, 12).$$

All of these are legal Nim moves.   ◀

Problem 159 *Compute* $13 \oplus 12 \oplus 8$.   Answer □

Problem 160 *Solve for an integer x so that* $38 \oplus x = 25$.   Answer □

Problem 161 *What is* $n \oplus n \oplus n \oplus \cdots \oplus n$?   Answer □

Problem 162 *Is the collection of nonnegative numbers with the operation* $\oplus$ *a group? (The notion of a group is defined later on in Section 4.9.)*   Answer □

Problem 163 *Try the method of Example 3.10.4 on the game* $(3, 10, 12)$. *Compute*

$$3 \oplus 10 \oplus 12 = (3 \oplus 10) \oplus 12 = 9 \oplus 12 = 5.$$

So are these the balancing moves

$$(3, 10, 12) \rightarrow (3 \oplus 5, 10, 12)$$
$$(3, 10, 12) \rightarrow (3, 10 \oplus 5, 12)$$

and

$$(3, 10, 12) \rightarrow (3, 10, 12 \oplus 5)?$$

     Answer □

Problem 164 *Try the method of Example 3.10.5 on the game* $(3, 10, 12)$. *Are these the balancing moves:*

$$(3, 10, 12) \rightarrow (3, 10, [3 \oplus 10])$$
$$(3, 10, 12) \rightarrow (3, [3 \oplus 12], 12)$$

and

$$(3, 10, 12) \rightarrow ([10 \oplus 12], 10, 12)?$$

     Answer □

3.10.8 Obituary notice of Charles L. Bouton

The obituary notices of Bouton at the time of his death in 1922 praised much of his academic work but made no mention of his solution of Nim. A century later we can see that he should be credited as one of the founders of combinatorial game theory. And Nim, at first seen as a particular example of an interesting game, turned out to be fundamental to the whole theory. His name now is far more likely to be mentioned in the context of game theory than the study of transformation groups that would have been his main interest during his career. As a tribute to him we include here this obituary notice (even though Nim does not appear) and, in our appendix, we include a copy of Bouton's paper on Nim.

CHARLES LEONARD BOUTON

Professor Charles Leonard Bouton died on February 20, 1922. See this BULLETIN, vol. 28, p. 82 (Jan.–Feb., 1922).

**A MINUTE READ BEFORE THE FACULTY OF HARVARD
UNIVERSITY**
March 28, 1922

Charles Leonard Bouton was born in St. Louis, Missouri, April 25, 1869. His father, William Bouton, was of Huguenot descent, and the family was long established in New England. At the close of the Civil War, William Bouton settled in St. Louis, where his regiment had been disbanded. Charles's mother, Mary Rothery Conklin, was also of old American stock; her grandparents were Scotch. William Bouton was an engineer by profession. His grandfather is said to have been the projector of the Erie Railroad, and was the author of the first article on its construction. Charles was the only one of the four sons who did not follow in his father's footsteps. The home atmosphere was academic and intellectually stimulating.

Bouton received his early education in the public schools of St. Louis, and took his first degree, that of Master of Science, at Washington University in 1891. Here, he came under the instruction of a highly skilled teacher of descriptive geometry, Dr. Edmund Arthur Engler. The next two years were given to teaching in Smith Academy, St. Louis, and these were followed by a year as instructor in Washington University, part of his work being to assist Professor Henry S. Pritchett. His next, and as it turned out, his last move was to Harvard. The years '94–'95 and '95–'96 were spent in the Graduate School. He took the master's degree at the end of the first year, and at the end of the second he was awarded a Parker Fellowship for study abroad. His two years at Leipzig were most profitably spent. He chose as his master that most original geometer, Sophus Lie, then at the height of his fame. As a matter of fact, Bouton was one of the great Norwegian's last pupils, for Lie returned to Norway in 1898 and died soon after. All of Bouton's subsequent scientific work bore the clear impress of Lie's genius. His two advanced courses, which he originated soon after his return to Harvard, dealt respectively with the theory of geometrical transformations and the application of transformation groups to the solution of differential equations. The graduate students who subsequently had the good fortune to prepare for the doctorate under his care generally took up subjects connected with the theory of transformations.

After receiving the doctorate at Leipzig in 1898 Bouton returned to Harvard and began a long period of work, broken only by occasional sabbatical absence. He threw himself with the greatest zeal into his duties as a teacher. At one time or another, beside the alternating advanced courses mentioned, he taught nearly every one of the lower and middle group courses in mathematics. No pains were too great for him to spend, either on the preparation of lectures or on helping the individual student, whether a Freshman or a candidate for the doctor's degree. His characteristic quality of scientific sanity was invaluable, for it led him always to emphasize that which was permanently important, and to avoid tinsel and sham. A fine example of his didactic sense is seen in a collection of problems on the construction of Riemann's surfaces, published in volume 12 (1898) of the ANNALS OF MATHEMATICS. He was equally successful in arousing the interest of a beginner by showing him a model or a diagram or an enlightening example of a new theory, and in guiding a graduate with sure hand toward researches of permanent value and importance.

Those qualities which made Bouton an admirable teacher were conspicuous in his other professional activities. He was an editor of the BULLETIN OF THE AMERICAN MATHEMATICAL SOCIETY from 1900 to 1902, and an associate editor of the TRANSACTIONS of the same society from 1902 to 1911. His power of keen yet kindly criticism, and his unerring mathematical judgment made him an efficient referee. His advice was prized by all who knew him, his opinion was always heard with respect, and his sanity was no less remarkable than his unselfishness. All of these qualities were drawn upon in full measure in the autumn of 1918 when, almost overnight, he was called to organize the mathematical instruction of nearly a thousand men in the Students' Army Training Corps. He carried this work through with conspicuous success, and the leading teachers of mathematics in the schools of this community, who enthusiastically rallied to the support of Harvard and the nation in that crisis, found in him a helpful guide and an efficient administrator.

His home life was beautifully quiet and peaceful. In 1907 he married Mary Spencer of Baltimore, and she, with their three daughters, Elizabeth, Margaret, and Charlotte, survives him. Yet for some time before the end, long dark shadows were crossing his life. The persistent after-effects of a hurried operation for appendicitis seemed to sap his strength. Family cares and anxieties multiplied, reaching a crisis in 1918 with the death of his youngest child. His breakdown in 1921 seemed but the inevitable end toward which events had long been tending. His death deprived the university of a faithful servant, and the community of a single-minded and upright gentleman.

From the Bulletin of the American Mathematical Society, 1922.

3.11   Answers to problems

Problem 76, page 97

This is quite easy since we can find a winning strategy for player II. In the first two moves there is a simple way of ensuring that the end position can never be white. In this case we proved the existence of a winning strategy for one of the players by specifying what it should be.

Problem 77, page 97

Suppose that player I does not have a winning strategy. What would this mean? Player I moves. Since he has no winning strategy, there is at least one move that player II can make that does not ensure a win for player I. So she should make that move. Then Player I moves again. Since he has no winning strategy, there is at least one next move that player II can make that does not ensure a win for player I. So she should make that move. This continues until the game is over and player II has won. That is her strategy. We know only that at each stage there must have been some correct strategic choice, but we do not know without detailed analysis what that move is.

Problem 78, page 97

Define an end position to be white if it is a win for player I or if it is a draw. Define an end position to be black if it is a win for player II. Then if we apply Problem 77 we know immediately that either player I has a strategy that must end in either a win or a draw or else player II has a wining strategy.

   We know from experience that player II has no winning strategy otherwise we would surely have found it before we were eight years old. We also know that there is no possible advantage in this game to going second. We can prove this, however, by a *strategy stealing argument*. We imagine that player II does have a winning strategy and we ask her to write it down. Then we steal it. If that strategy did work we could use it to win ourselves. Make a first random move. Then follow the stolen strategy as if you were player II (placing X's where the strategy tells you to place O's). If the strategy requires you to place a mark on a square that you previously used, just make a new random move. The strategy guarantees a win. But it can't because player II should always win with correct play. Thus there is no winning strategy for player II as we suspected.

Problem 79, page 97

The fact of two games being "identical" is important to our investigations. For this game arrange the nine numbers 2–10 in a 3×3 square array so that the sum along any row or column or diagonal is exactly 18. Figure 3.29 illustrates this.

3	10	5
8	6	4
7	2	9

Figure 3.29: The game of 18 is identical to tic-tac-toe

Then a move in the game of 18 for a player consists essentially of choosing a position in the array and marking it with either an X or an O depending on whether he is the first or second to move. The two games are then easily checked to be identical.

After a child has mastered the game of tic-tac-toe it would be a good exercise to have them play this game. At some point they will spot the strategy (assuming the arithmetic skills are relatively strong) and perhaps even notice that the game is equivalent to tic-tac-toe.

Problem 80, page 97

This game too is the same as a tic-tac-toe game.

For this game arrange the nine cards in a 3×3 square array so that the rows, columns and diagonals are the same as the eight winning card combinations. Figure 3.30 illustrates this.

$J\diamondsuit$	$Q\diamondsuit$	$K\diamondsuit$
$J\spadesuit$	$Q\spadesuit$	$K\spadesuit$
$J\heartsuit$	$Q\heartsuit$	$K\heartsuit$

Figure 3.30: The card game is identical to tic-tac-toe

The game has the appearance of being a typical card game because the winning combinations are rather familiar ones, but it is nothing more than the usual trivial game of tic-tac-toe described in different language.

Problem 81, page 99

The full strategy is described in Section 3.2.5. At this stage you may not be able to articulate the strategy in the same language that we will use, but you can experiment enough with the game to devise a way of winning. As before, you should discover that there are precisely two kinds of positions: ones in which we can make a good move and ones in which no good move can be made.

Start with the simplest positions and determine which ones can be classified as good (or winning) positions and which positions are bad (or losing).

Problem 82, page 102

The final position is balanced because $1+2=3$. For a balanced position $A+B=C$, then no matter what move is made, one side of the equation is reduced while the other remains the same. Thus, any move will destroy the balance.

For an unbalanced position $A+B \neq C$. There are two cases, $A+B < C$ and $A+B > C$. If $A+B < C$, we can reduce C so that $A+B = C$. If $A+B > C$, then $A > C-B$ so we can reduce A to re-establish the balance. In either case, there will be a move to re-establish the balancing equation $A+B=C$.

Problem 83, page 102

Call those positions in the four-marker game which satisfy the equation $D-C = B-A$, *balanced* and all other positions for which $D-C \neq B-A$ *unbalanced*. Make sure to verify that the three conditions for balance are met.

For example, if a position is balanced, then we need to show that every immediately following position is unbalanced. A move requires us to change the position of exactly one of the four markers. Clearly any such move will change one side of the equation

$$D-C = B-A$$

and produce an unbalanced position.

On the other hand if a position is unbalanced then

$$D-C \neq B-A.$$

In that case either

$$D-C > B-A \quad \text{or else} \quad D-C < B-A.$$

Which marker would you move in each of these two cases?

Problem 84, page 102

Not so easy. In fact the discussion so far might lead you to believe that you should search for just the right equation, similar to the situation for the three and four-marker games. This does not work.

If you run out of ideas (as we fully expect you will) move on to the next section and read about some other combinatorial games. We will return to this problem later with some fresh ideas.

Problem 85, page 105

This exercise is an essential one to perform in order to see how the balancing definition works. To study a game this way one needs only to know, for any given position, all of the positions which follow from it by a single legal move. For 2–pile Nim this is easy.

Describe a position in the game as (m, n) if there are m sticks in the first pile and n sticks in the second. Try to compute the balancing number for $(2, 1)$ for example. List all of the positions which follow directly from $(2, 1)$:

$$(2, 1) \rightsquigarrow (1, 1), (0, 1), \text{ and } (2, 0).$$

That means you cannot compute the balancing number for $(2, 1)$ until you know the balancing number for each of these other positions.

Start at the bottom (i.e., the end of the game). If you compute the balancing numbers in the order suggested in Figure 3.31 the definition is easy to apply. Here we start at the lowest depth (the end position) and work back to higher depths a step at a time. Note that, if you have found all the balancing numbers at any depth, you will be able to find all the balancing numbers at the next higher depth.

Position	Depth	Balancing Number	Position	Depth	Balancing Number
(0,0)	0	0 [balanced]	(5,0)	5	1 [unbalanced]
(1,0)	1	1 [unbalanced]	(0,5)	5	1 [unbalanced]
(0,1)	1	1 [unbalanced]	(3,3)	6	0 [balanced]
(1,1)	2	0 [balanced]	(4,2)	6	1 [unbalanced]
(2,0)	2	1 [unbalanced]	(2,4)	6	1 [unbalanced]
(0,2)	2	1 [unbalanced]	(5,1)	6	1 [unbalanced]
(2,1)	3	1 [unbalanced]	(1,5)	6	1 [unbalanced]
(1,2)	3	1 [unbalanced]	(0,6)	6	1 [unbalanced]
(3,0)	3	1 [unbalanced]	(6,0)	6	1 [unbalanced]
(0,3)	3	1 [unbalanced]	(6,1)	7	1 [unbalanced]
(2,2)	4	0 [balanced]	(1,6)	7	1 [unbalanced]
(3,1)	4	1 [unbalanced]	(5,2)	7	1 [unbalanced]
(1,3)	4	1 [unbalanced]	(2,5)	7	1 [unbalanced]
(4,0)	4	1 [unbalanced]	(4,3)	7	1 [unbalanced]
(0,4)	4	1 [unbalanced]	(3,4)	7	1 [unbalanced]
(3,2)	5	1 [unbalanced]	(4,4)	8	0 [balanced]
(2,3)	5	1 [unbalanced]	(5,3)	8	1 [unbalanced]
(4,1)	5	1 [unbalanced]	(3,5)	8	1 [unbalanced]
(1,4)	5	1 [unbalanced]	(6,2)	8	1 [unbalanced]

Figure 3.31: Balancing numbers for 2–pile Nim.

For example we can illustrate with the position $(3, 1)$ at depth 4. The possible moves from this position are:

$$(3, 1) \rightsquigarrow (2, 1), (1, 1), (0, 1), \text{ and } (3, 0).$$

From the table we already know the balancing numbers for these positions are

$$1, \ 0, \ 1, \ \text{and} \ 1.$$

Consequently, since 0 appears in this list,

$$\text{Balance}(3,1) = 1.$$

You can continue much further if you are not yet bored. Applying the definition, even in such a simple case as 2–pile Nim, can be quie tedious. At some point you can spot the pattern and can figure out a correct strategy for play.

How could we use such a table? Well, if we simply cannot spot the pattern, then make a large table. While you are playing consult the table. For example, you are at the position $(4,2)$ in the game and must decide to make a move. In the table you see that it is unbalanced. Therefore there must be a move that you can make to rebalance it. Look for a balanced position above it in the table and see if you can move to that position. Above $(4,2)$ in the table are the balanced positions $(2,2)$, $(1,1)$, and $(0,0)$. The only one you can reach is $(2,2)$. Accordingly then you make your move: take two sticks from the first pile.

The other obvious way to use such a table is to spot patterns. It is pretty clear from the table so far that the only balanced positions are those of the form (m,m). Any position (m,n) with $m \neq n$ appears likely to be unbalanced. How would you go about proving this?

Problem 86, page 105

We recognize these three statements as the same as those for the marker games in Statement 3.2.2. It is clear that a player can always win from a black position by choosing to move to a red position. But if a black position is balanced, there is no strategy that will always win from that starting point because your opponent can always produce a balanced position. Thus black positions must be unbalanced. For much the same reason red positions must be balanced.

This red and black argument thus allows us to find all balanced and unbalanced positions without going through the computations involved in finding the value of $\text{Balance}(p)$ for every position in the game. If we can spot a pattern that follows this red/black scheme we can immediately claim to have found all the balanced positions.

Inductive proof   You might also wish to prove that all red positions are balanced by induction. Start at depth zero. These positions are red and are balanced. At depth one all positions are unbalanced and these must be black since they move only to the depth zero positions that are red. Assume that red=balanced and black=unbalanced for all depths $0,1,2,\ldots n-1$ and show that the same must be true at depth n.

Problem 87, page 105

The conjecture is that a position (m,n) is balanced in 2–pile Nim if and only if $m = n$. Define a position (m,n) to be red if $m = n$ and to be black if $m \neq n$. Just check that

1. The end position $(0,0)$ is red.

2. Any red position (n,n) can move only to a black.

3. From any black position (m,n) with $m \neq n$ there is at least one move to a red position.

Then apply the red/black argument to conclude that red positions are balanced and black positions are unbalanced. Note that, although we used the balancing numbers to guide us towards our conjecture, we do not need them for our proof that a position (m,n) is balanced if and only if $m = n$.

Problem 88, page 105

The name SNIM is meant to suggest "Stupid" Nim. If you play the game with a friend you will see why. For example, start with the position $(3,5)$. A balancing move is to produce $(3,3)$. Your opponent will know where this is heading and add a stick to one of the piles. The game play looks like this:

$$(3,5) \rightsquigarrow (3,3) \rightsquigarrow (3,4) \rightsquigarrow (3,3) \rightsquigarrow (4,3) \rightsquigarrow (3,3) \rightsquigarrow (4,3) \rightsquigarrow \ldots$$

and continues forever, with no end and no winner.

It is an essential part of our theory that the game we are studying must be finite. It is possible to study infinite games, but the strategy that is based on the balanced and unbalanced analysis depends on the game being finite.

Problem 89, page 105

Declare those positions in the four-marker game that satisfy the equation

$$D - C = B - A$$

to be red and declare the others to be black. Now just check that the three conditions hold. Note that this is precisely what we did before in Problem 83. Now we have defined what balanced and unbalanced actually mean for any game. An appeal to the red/black argument shows that, indeed, the positions for which the markers satisfy the equation $D - C = B - A$ are the balanced positions in the sense of our new definition of balanced.

Problem 90, page 105

Such a game is not a game of strategy, although it might well appear to the players to involve some mysterious strategy. The main feature of such a game is that no strategy or effort is needed to play the game. No matter what the players do the winner of the game is determined from the very first move. If Player I starts with a balanced position he loses. If Player I starts with an unbalanced position he wins no matter how he chooses to move. We have seen such a game already in Section 2.6.3 when we were studying triangulations of polygons.

Problem 91, page 105

The answer is no, if Player I is skilled at finding the balancing moves. If Player II suspects that Player I is not quite so skilled then there is an obvious, but weak, strategy. She should always select a move that leaves the position as complicated as possible in the hope that her opponent will make a mistake.

Problem 92, page 106

The balanced positions are 5, 10, 15, 20, ... (any multiple of 5).

Problem 93, page 106

One way to calculate balanced positions is to create a *sieve* for this game. Here is how that works. First we note that 0 is a balanced position—the player facing this position has no move. Thus for any number s that is a square, $s = 1, 4, 16, 25, \ldots$, the position $0 + s$ is unbalanced.

The player facing such a position may take all the sticks leaving 0, which is a balanced position. The smallest number not yet shown to be unbalanced is 2. This number is balanced. (Check this). Thus for any square s, the position $2 + s$ is unbalanced, so 3, 6, 11, 18, ... are all unbalanced. The smallest number not yet shown to be unbalanced is 5 . Thus 5 is balanced.

Continuing in this way we find many unbalanced positions and see that the only balanced positions less than 25 are 0, 2, 5, 7, 10, 12, 15, 17, 20, and 22.

Does this inspire you to predict[4] all the unbalanced positions over 25? The pattern emerging might suggest that we can obtain all balanced positions by alternating adding a 2 and a 3 to the previous balanced position found. We can do this *ad infinitum*. Does that work? (See Problem 94 and Problem 95 before jumping to any conclusions.)

---

[4]If an I.Q. test were to ask for the next terms in the sequence 0, 2, 5, 7, 10, 12, 15, 17, 20, 22, ... what would most people respond?

Problem 94, page 106

The sieve method that we used in solving Problem 93 led to the balanced positions $0, 2, 5, 7, 10, 12, 15, 17, 20$, and 22.

We can apply the sieve again and keep going to obtain all the remaining balanced positions less than 100. Or we can just claim to see the pattern: all balanced positions seem to obtained by alternating adding a 2 and a 3 to the previous balanced position found.

Whoops! Continuing our sieve process we find, instead, that the remaining balanced positions less than 100 are 34, 39, 44, 62, 65, 67, 72, 77, 82, 85, and 95. We no longer see this same 2 and 3 pattern, nor indeed any pattern. Do you?

Problem 95, page 106

We have no idea what type of formula might work. In fact, as of July 2010 no one had found one. Computers were put to work generating balanced positions. They found that in the first 40 million positive integers, about 180000 were balanced positions. They had a number of conjectures about how often balanced numbers had various numerals in the units position. Only one such number, 11356, had a 6 in that position. How their conjectures turn out remains to be seen. Many references to this problem can be found by checking Wikipedia. Look under "subtraction games."

This is an example of a problem that looked simple at first but proved difficult. It is also an example of a process that after many initial computations suggested patterns [as we found in solving Problem 93 up to 25], only to be proved false. We discuss more extreme examples of this in a later chapter in Volume 2 where we discuss the famous (incorrect) conjectures of Polya and Mertens.

Problem 96, page 106

If the subtraction set is
$$S = \{1, 2, 3, 4, 5, 6, 7, 8, 9, 10\}$$
then the balanced positions are the multiples of 11.

Problem 98, page 108

The one pile and two pile games would likely have caused no troubles. The three pile game is much more difficult, but maybe you spotted exactly what the balanced and unbalanced positions are. Later on we will find the exact strategy for the coin game by comparing it to the binary bits game.

Problem 99, page 108

This observation is important to make. We cannot use our balanced and unbalanced arguments unless the game is finite.

It is not true that each move of the game reduces the number of coins on the table. It is occasionally possible to add more coins than are removed. But each move of the game does reduce the total value of all the coins in play (check this). Thus the total value goes down with each move and eventually reaches zero when there are no more coins in play and no further moves possible.

Problem 100, page 108

Begin by displaying the cards on the table in a 13×4 rectangular array. The bottom row displays all the 2's (if there are any) and so on up to the top row that displays the aces. Then recognize that the display can be translated to binary bits with no loss of information.

As we did for the coin game we use the bit 1 for YES and the bit 0 for NO to indicate whether a card is or is not on the table. Again we can simplify the moves in the game if we realize that removing a card simply changes a YES to a NO, i.e., it changes a 1 bit to a 0 bit. Similarly adding a card changes a NO to a YES, i.e., it changes a 0 bit to a 1 bit. Once again we are just flipping bits instead of playing with cards.

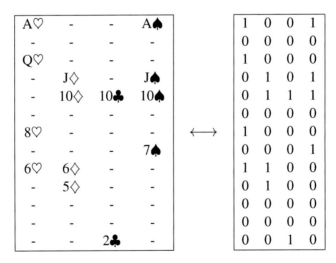

Figure 3.32: A position in the card game.

Figure 3.32 shows such a display, along with the equivalent array of binary bits, for a game in which the deal is

$$A\heartsuit, \ Q\heartsuit, \ 8\heartsuit, \ 6\heartsuit, \ J\diamondsuit, \ 10\diamondsuit, \ 6\diamondsuit, \ 5\diamondsuit, \ 10\clubsuit, \ 2\clubsuit, \ A\spadesuit, \ J\spadesuit, \ 10\spadesuit, \ 7\spadesuit.$$

The play of the game is exactly the same as the play of the game of binary bits so the game is completely analyzed by that equivalent game.

If you do wish to play this game perhaps you might want to reduce the size of the game by using only part of the full deck of cards. Otherwise the play of the game may take a long time. But do notice that the game is easy to play.

For example *after you have discovered the strategy for the game* you would instantly recognize that the position in Figure 3.32 is unbalanced and that the only balancing move is to take away $Q\heartsuit$ and $8\heartsuit$ and to add to the table the cards $10\heartsuit$, $7\heartsuit$, $5\heartsuit$, and $2\heartsuit$. This will take only seconds to spot.

Problem 101, page 110

Usually for our games this is obvious. Here you might have been bothered by the fact that a play of the game removes one binary bit but could well add many more. However, you might have noticed that only finitely many positions are possible, and no position can be repeated. (Why can none of the position be repeated?)

Another solution is don't simply count binary bits, but do a *weighted* count. Each 1 bit on the bottom row receives a weight of 1. Each 1 bit on the second row receives a weight of 2. Each 1 bit on the third row receives a weight of 4, and so on. Now we see that every play of the game, while it may not reduce the actual count of 1 bits, it does reduce the weighted count. When the weighted count is zero there are no more 1 bits and the game stops.

Problem 102, page 110

You can start with a 2×2 game. There are only a few possibilities and you should notice the pattern. By the time you have mastered the 3×2 game the strategy is apparent.

The mirror strategy If the two columns are identical then the position is balanced. Thus a balanced position has either two 1 bits in each row or two 0 bits in each row. If player I makes a move in such a position then player II just *mirrors* the same move back at him in the other column. Eventually she wins.

If the two columns are not identical the position is unbalanced. If a player can make a move in such a position he just balances the game by making the two columns identical. Then the next player is doomed since every move she makes unbalances the position.

The mirror strategy plays an important, strategic role in a number of games that have this feature: the game can be split into two identical pieces. Some authors prefer to call it the *Tweedledum-Tweedledee strategy*. Whatever Tweedledum does in one of the columns, Tweedledee does the same in the other column. Tweedledee wins.

Problem 103, page 110

The ones with an odd number of 1 bits are unbalanced. The one with an even number of 1 bits is balanced. These positions are very close to the end of the game and it is always easy to determine in such cases which positions are balanced or unbalanced.

Problem 104, page 110

Just play the games and see if you can force a win or not. All of these are unbalanced.

Problem 105, page 110

All of these are balanced.

Problem 106, page 110

In Problem 104 all positions were unbalanced and all these positions had one or more odd rows, i.e., rows with an odd number of 1 bits. In Problem 105 all were balanced and all these positions had only even rows—every row had an even number of 1 bits. Now do you have a conjecture?

Problem 107, page 111

Note that this is the same scheme that we use in a red and black argument, although we have expressed it in the even and odd language. The first two statements are quite clear. In the end position there are only zeros so certainly that is an even position. If the player starts with an even position he must select a 1 bit in some column to change. At that point he has already produced a row with an odd number of 1 bits and so an odd position.

Let us check the final statement. If the position is odd then then there are one or more rows with an odd number of bits. Take the topmost odd row and choose a 1 bit to change. That makes that row now even. But there are possibly other odd rows, each of them lower than the one you chose. Each of those rows can be adjusted by changing the bits as necessary. The result is an even position.

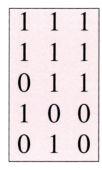

Figure 3.33: An odd position.

For example, Figure 3.33 illustrates an odd position in a 5×3 game. It is odd because four of the rows have an odd number of 1 bits. To balance will require that we change all four of these rows (but leave the one even row alone). It is easy and obvious how to do this. Figure 3.34 shows one way, but there are two other ways in which you could have changed this position to an even one using a legal move in the binary bits game.

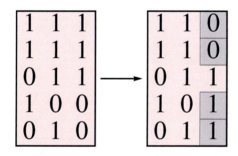

Figure 3.34: How to change an odd position to an even position.

Problem 108, page 111

It is clear that even and odd positions behave precisely as do balanced and unbalanced positions. The strategy of the game that will work is to start (one hopes) with an odd position. There is a move that will change it to an even position. Your opponents move will undoubtedly change it back to an odd position. This continues until the game stops and we know it stops at an even position. It must have been you that made the last move and so you win. (Of course, if you must start with an even position just play modestly hoping that your opponent will make a mistake and leave an odd position.)

Problem 109, page 111

It is not difficult to see that the argument of Problem 107 applies to any size game of binary bits. Thus the analysis of the game in terms of even and odd rows will solve the the the $m \times 4$ game, the $m \times 5$ game and indeed the $m \times n$ game.

Problem 110, page 111

You shouldn't have to convert to binary, but you should be able to spot the correct strategy. There are an odd number of dimes (but an even number of quarters). So you must remove one of the three dimes. You have to balance the nickels too, but the pennies are balanced.

Notice, that with this strategy, the game is easy to play. Against a player who does not know the correct strategy you win every time and your game play is very rapid. Unfortunately a shrewd opponent might be able to spot what you are doing and realize how doomed he is each time he faces a position with an even number of coins of each type.

Problem 111, page 112

It makes it messier and, perhaps, more confusing for your opponent. But if you work on it for a while you will see that this game is exactly equivalent to binary bits too and is played with the same strategy.

Problem 112, page 112

A winner would still be declared when the last coin is removed, but any person playing the game would prefer to be a loser and walk away with the most money. Thus the right strategy is to select the pile of largest value at each turn and take all the money. In the end you lose the game and leave richer.

In the language of game theory we have essentially changed the game to a *scoring game*. Many card games do not end with the winner the player making the last move, but the player who accumulates the most points. Our theory of balanced and unbalanced positions does not apply to scoring games.

Problem 113, page 112

It is a good choice of game to impose on a friend who considers himself bad with arithmetic. It appears to require great skill in working with numbers, but this is deceptive. The structure of the game play is simpler than it at first appears: some non-zero numbers are merely replaced by zeros.

As soon as this occurs to us we realize that the game is not just "similar" to the game of binary bits; it is identical. After a few plays of the game we recognize that all that matters is whether a number is zero or non-zero. Replace all the non-zero numbers with the binary bit 1. Then the rules of the game are identical to those for binary bits.

To play this game just convert any position to the equivalent position in binary bits and play the strategy that we have described.

$$
\begin{array}{rrr}
10 & -9 & 0 \\
-3 & 11 & -32 \\
0 & 11 & 32 \\
4 & 0 & 0 \\
0 & -14 & 0
\end{array}
$$

Figure 3.35: A position in the numbers game.

For example, Figure 3.35 shows a position in 5×3 numbers game. We must determine whether the position is unbalanced or balanced. If it is balanced then we must find at least one move that will balance it.

Figure 3.36 illustrates how we can solve this problem by converting that position to a position in a binary bits game. We make the correct balancing move in the bits game, and then return back to an equivalent position in the numbers game.

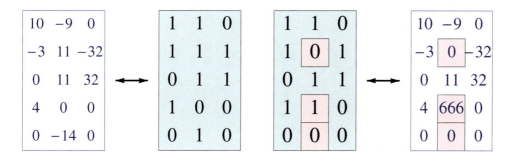

Figure 3.36: Playing the numbers game.

The choice of 666 is of course arbitrary here and intended only to irritate an opponent; any nonzero number will do the trick. Here we see that, while the game had a different appearance to the game of binary bits, it has exactly the same structure—the two games are equivalent once we find how to match up positions and moves. Sometimes this is easy to see, sometimes not.

Problem 114, page 112

It is a good choice of game for a child who needs some practice on the order of letters in the alphabet. The structure of the game suggests something tricky about words and letters but the game is completely equivalent to the binary bits game.

Begin by displaying the letters in each of the n words in a $26 \times n$ rectangular array. The bottom row displays all the a's (if there are any) and so on up to the top row that displays the z's. Then recognize that the display can be translated to binary bits with no loss of information.

As we did for the coin game and the card game we use the bit 1 for YES and the bit 0 for NO to indicate whether a letter is or is not in the word that corresponds to a column. Again we can simplify the moves in the game if we realize that removing a letter simply changes a YES to a NO, i.e., it changes a 1 bit to a 0 bit. Similarly, adding a letter (if that letter was not already there) changes a NO to a YES, i.e., it changes a 0 bit to a 1 bit. Once again we are just flipping bits instead of playing with words.

For example the position

$$[\text{ ebbde caecde cddc}]$$

in this game can be displayed as the 5×3 arrays of bits in Figure 3.37.

$$
\begin{array}{ccc}
1 & 1 & 0 \\
1 & 1 & 1 \\
0 & 1 & 1 \\
1 & 0 & 0 \\
0 & 1 & 0
\end{array}
$$

Figure 3.37: A position in the word game.

Since no letters higher than "e" appear we do not need any higher rows. This is an unbalanced position and can be easily balanced in the manner shown in Figure 3.38.

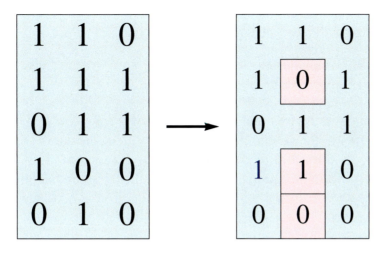

Figure 3.38: Balancing that same position in the word game.

Thus a correct response in this position would be the play

[**ebbde caecde cddc**] ⤳ [**ebbde cebe cddc**]

changing just the second word. While the game had a different appearance to the game of binary bits, it has exactly the same structure—the two games are equivalent once we find how to match up positions and moves. (Indeed, the position in this game that gave rise to the scheme in Figure 3.37 is exactly the equivalent position that we saw before in the numbers game play of Figure 3.36.) As always, sometimes this is easy to see, sometimes not.

Problem 115, page 112

Maybe so, maybe not.

Problem 116, page 114

Remove all sticks. You win. A balanced position contains no sticks; every pile that contains one or more sticks is unbalanced.

Problem 117, page 114

This involves experimenting until you see what is involved, formulating a conjecture of what are the balanced positions, and then verifying that the three conditions required for the set of balanced positions are met. In other words, you must show that your conjectured set of balanced positions meets the three conditions of Statement 3.2.2:

1. The final position (no sticks remaining) is balanced.

2. If a position is balanced, then no matter what move our opponent makes, the resulting position is unbalanced, and

3. If a position is unbalanced, then there is a move we can make which results in a balanced position.

Define a position in the two-pile game as (m,n) if there are m sticks in the first pile and n sticks in the second pile. If, near the end of the game you leave your opponent $(1,1)$, you will evidently win. If he leaves you $(1,0)$, $(2,0)$, $(3,0)$ etc. you will win immediately by taking all of the sticks in that pile.

You can easily verify that a position (m,n) should be called balanced if $m = n$ and unbalanced if $m \neq n$. Check the three conditions.

We have already seen this situation in our solution of Problem 102 in the game of binary bits. Let us repeat what we learned there but modified now to discuss 2-pile Nim. This will save the reader some flipping.

The mirror strategy   The balanced position (m,m) in the 2-pile Nim game offers the player a chance to use the mirror strategy. If player I makes a move in such a position then player II just *mirrors* the same move back at him in the other pile. Eventually she wins. The mirror strategy (or Tweedledum-Tweedledee strategy) we have seen before. Whenever a game can be split into two identical "subgames" this strategy will be successful. Whatever Tweedledum does in one of the piles, Tweedledee does the same in the other pile. Tweedledee wins.

Problem 118, page 114

Think of the game (m,n,m,n) as being two identical games of 2–pile Nim by placing a mirror in the middle:

$$(m,n \mid m,n).$$

Now your opponent makes a move on one side of the mirror and you just repeat it on the other side. Since you always leave a position that has this mirror symmetry you must be the winner as the final position $(0,0,0,0)$ has this same symmetry.

This shows that every position (a,b,a,b) in the game is balanced, but it does not find *all* balanced positions in 4–pile Nim. This does not matter to us because the mirror strategy allows us to control the game and avoid encountering positions that we do not know how to balance.

Since the mirror strategy is so easy to apply, it is very seductive. You might think for a while that it will help in all Nim games, but this is not so. We will need some fresh ideas even for 3–pile Nim.

Problem 119, page 114

For the first move of the game take away one or two coins so as to leave two separated rows containing the same number of coins. For example (as illustrated in Figure 3.14) if there are 14 coins and we would remove the two middle coins so as to produce two separate games of Kayles with 6 coins in each game.

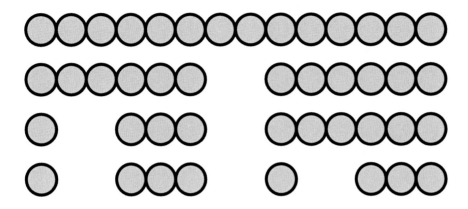

Figure 3.39: A sequence of moves in a game of Kayles.

Apply the Tweedledum-Tweedledee strategy (i.e., the mirror strategy) to all subsequent moves. Whatever your opponent does to one side you respond with the same thing on the other side. You win.

There is an odd thing about this strategy, apart from the fact that we must have the first move in order to apply it. We do not know all possible balanced and unbalanced positions and yet we can win by controlling the flow of the game to visit only positions that allow us to apply the mirror strategy. The starting position is always unbalanced, every move we make is a balancing move, and every move our opponent makes is necessarily an unbalancing move.

If you wish to apply this in practice, note that you will surely win every time you start first, and that you may well win occasionally when you start second. That makes the game quite favorable to you. But the strategy is always the same and your opponent may spot what you are doing. To avoid this try to destroy the symmetry a bit by using a variety of coins. Figure 3.40 shows the same sequence of moves, but here the coins are varied and this little slight-of-hand trick helps obscure what might have been an obvious strategy.

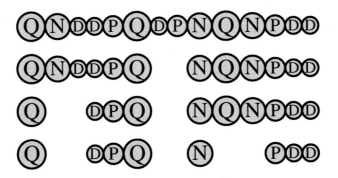

Figure 3.40: The same sequence of moves in a game of Kayles.

Problem 120, page 115

The opening position in a game of Kayles where the coins are arranged in a straight line is always unbalanced. Thus Player I can always win. If the coins are arranged in a circle then the first move must break the circle and the coins are (essentially) back to being arranged in a straight line.

Thus the opening position in circular Kayles is always balanced and so Player II will win simply by waiting until the second move and playing the usual Kayles strategy.

Problem 121, page 115

It is easy to see that $(1,2,2)$ is unbalanced, because it leads directly to the balanced position $(0,2,2)$ (the same as the 2–pile game $(2,2)$ which we saw was balanced). The game $(1,1,2)$ is unbalanced for the same reason.

Problem 122, page 115

One argument is the working-backwards one. The position $(1,2,3)$ is not far from the bottom of the game. The positions that follow from here are

$$(1,0,3),(0,2,3),(1,1,3),(1,2,2),(1,2,1), \text{ and } (1,2,0).$$

If all of these are unbalanced then $(1,2,3)$ must be balanced. If one of these is balanced then $(1,2,3)$ is unbalanced. If we don't know the status of one of these then do the same thing to that position to find out what is the situation.

Another argument is to show how we could respond to every move by our opponent in a way to produce a balanced position. That proves that all of the moves directly from the $(1,2,3)$ position must themselves be unbalanced.

Let us give the details by this method to show that the game $(1,2,3)$ is balanced. This will give us an indication of what one might do in order to show that a position is balanced. Note how the little bit of knowledge we have already obtained simplifies our task considerably.

We must show that no matter what move our opponent makes from the position $(1, 2, 3)$, we can find an answer which leaves a balanced position. The chart below shows our answer to each of the six allowable moves for our opponent.

Position after opponent's move	Position after our answer
(0,2,3)	(0,2,2)
(1,1,3)	(1,1,0)
(1,0,3)	(1,0,1)
(1,2,2)	(0,2,2)
(1,2,1)	(1,0,1)
(1,2,0)	(1,1,0)

Figure 3.41: Positions in the game $(1, 2, 3)$.

Thus, no matter what our opponent did, we were able to make a move which left our opponent with a balanced position. We know each of these positions is balanced because they are all equivalent to 2–pile Nim games of the form (n, n), and we have already seen that every such position is balanced.

Problem 123, page 116

All of the games are balanced.

Problem 124, page 116

All of the games are unbalanced.

Problem 125, page 117

All of the games are unbalanced.

Problem 126, page 117

All of the games are unbalanced.

Problem 127, page 117

Problem 127 carries the key to the whole structure of Nim. It will certainly be worthwhile to one's understanding of Nim to put in whatever time is necessary to discover this key!

Problem 128, page 125

If the highest row with an odd number of 1's in it has one 1, there will be exactly one balancing move, and that move must be made from that pile which corresponds to that column. If it has three one's in it, there will be three possible moves to balance the position. One such move will be possible from each column.

Problem 129, page 125

This game can be balanced in these three ways. Remove 3 from the first pile or remove 7 from the second pile or remove 11 from the third pile. While these are not obvious when the problem is expressed in decimal notion you should have little trouble if you express the problem in binary.

Problem 130, page 126

The answer is *no*. This is so because there are only three piles and there cannot be more than one balancing move from each pile.

This statement is correct since, if it were not, there would be two positions (a,b,c) and (a,b,d), both balanced, with c and d different. But this would imply that a move from a balanced position could result in another balanced position, an impossibility.

Problem 131, page 126

The largest number number of balancing moves is 9.

Problem 132, page 126

The largest number number of balancing moves is 11.

Problem 133, page 126

Without computing anything it is clear that the position is unbalanced and that the only move will be to take most or all of the big pile that contains 100000. That answers the question but you might want also to find out exactly how many to take.

Problem 134, page 126

Why is the Nim game

$$(136, 72, 48, 40)$$

unbalanced? Note that $136 > 2^7 > 72$. Thus if we were to convert this to a position in a game of binary bits the top row for this game has only one 1, and the game is unbalanced. A balancing move must come from the pile with 136 sticks.

Generally

$$(\text{really big}, \text{not so big}, \text{smaller still}, \dots)$$

makes it easy to spot which pile to choose and why the position is unbalanced.

Problem 135, page 126

Decline. If your opponent knows the strategy then you will surely lose. If you think your opponent is naive then start by taking one stick from the largest pile. If she plays a balancing move then she knows the strategy.

Another opening gambit is to offer politely that she should start instead. This appears very courteous since, in almost all games, it is an advantage to start. If she insists that you should start you can test her out this way: agree to start, but say "I'll start, but let's make it more interesting" and quickly add a bunch of sticks to form a new pile. Think for awhile before making the move: then remove all the sticks from the new pile and say "Your move."

This is really a great joke. If she is unamused then you know she is fully aware of the strategy, for you have immediately turned her into player one starting the original balanced game.

Problem 136, page 126

Nice game and more interesting than Nim since there is money on the table. It also appears to add a new element of strategy which increases the strategic interest.

But the deadly Nim strategy cannot be defeated by a player adding coins he has collected. Each time a player takes his coins and adds them to a pile in a balanced position, take those coins back and add to your own collection of coins, thus returning to a balanced position.

The game is a bit unusual for us in that, if neither player caught on to this strategy, the game could go on forever. But if one player adheres to the take-back strategy the game ends in a finite number of steps with a winner.

Problem 137, page 128

To balance an unbalanced 4–marker game requires two *gaps* be made equal. To balance an unbalanced 2–pile Nim game the *piles* must be made equal.

The similarity between the strategies for the 4–marker game and the 2–pile Nim game is this: the strategy in the first is to make the gap between A and B

the same as the gap between C and D. The strategy in the second is to make number of sticks in each pile the same.

Let us write the balanced positions for the 3-marker games in the form

$$A - 0 = C - B.$$

This is more conveniently written as

$$A - 1 = C - B - 1$$

since these two expressions actually measure the gap correctly. This translates into

the gap between A and 0 = the gap between C and B

or

# of empty holes to the left of A = # of empty holes between C and B.

For 2–pile Nim, we can formulate the balanced positions as

# of sticks in the first pile = # of sticks in the second pile.

Thus, we hope to be able to use the strategy for 2–pile Nim to give us a strategy for the 4–marker games. Let us see how that works.

A move in a balanced 4–marker game might be to move B or D. If B is moved, the gap between A and B is reduced. If D is moved, the gap between C and D is reduced. We can restore the balance by moving D or B, respectively, the same number of holes that our opponent moved B or D.

This is in perfect correspondence to balancing an unbalanced 2—pile Nim game. In this game our opponent takes sticks from one of the two equal piles. Our answer is to take the same number of sticks from the other pile, thus restoring the balance.

The other possible move in the 4–marker game consists of moving A or C. If A is moved the gap between A and B is *increased*. If C is moved the gap between C and D is *increased*. This does not correspond to a move in Nim, because in Nim the size of a pile must be *reduced*, not increased. In that case, however, we can move B or D to restore the gap to its original size.

Thus, our strategy has two parts to it: if our opponent reduces a gap, we reduce the other gap the same amount; this corresponds to Nim. If the opponent increases a gap, we reduce it the same amount.

Problem 138, page 128

The answer to Problem 137 (for the 4–marker game) will help. The main starting point is the equation $A + B = C$, rewritten as

$$A - 0 = C - B.$$

This shows that the two gaps between 0 and A and between B and C are to be the focus of the strategy.

Problem 139, page 128

For a 2-marker game there are two gaps: the gap between 0 and A and the gap between A and B. Only the second gap is of interest to us. The equation $B = A + 1$, rewritten as

$$B - A = 1$$

reveals that the balanced positions are the ones with that gap closed and, indeed, we do remember that the correct strategy is to balance the position by completely closing that gap. This is equivalent to a position in a 1–pile Nim game in which the only balancing move is to take at once all of the sticks from the pile.

Problem 140, page 128

Reread the material in this section until you fully understand it, and then work on Problem 140 without reading ahead to Section 3.6.2 which gives a full solution.

We were guided in the case of the 3–marker and 4–marker games by the equations

$$A - 0 = C - B \text{ and } B - A = D - C.$$

These told us exactly which gaps to work on and suggested a comparison with a 2–pile Nim game. Now we need to attempt to apply the same principle to the 5 and 6–marker games and seek a comparison to a 3–pile Nim game. Start by deciding on three gaps that you will use in your strategy.

Problem 141, page 131

The marker game with markers at

$$10, \ 15, \ 20, \ 25, \ 40, \ 50, \ 60 \text{ and } 80$$

corresponds to the Nim game (4, 4, 9, 19).

Problem 142, page 131

The marker game with markers at

$$10, \ 15, \ 20, \ 25, \ 40, \ 50 \text{ and } 60$$

corresponds to the 4–pile Nim game with gaps $(9, 4, 14, 9)$.

Problem 143, page 132

This corresponds to a 3–pile Nim game with the gap position at $(3, 0, 6)$. The only balancing move is to move from $(3, 0, 6)$ to $(3, 0, 3)$. This corresponds to moving the marker at 27 down to 24.

Problem 144, page 132

While most of the markers are at the same position in Problem 143 the gaps are completely different and the game has very different play. This marker game corresponds to a 3–pile Nim game with the gap position at $(4,3,5)$.

Problem 145, page 132

This corresponds to a 4–pile Nim game with the gap position at $(3,0,6,76)$. Without much thinking you should immediately move the marker at 100 down a long way. How long?

Problem 146, page 132

This corresponds to a 5–pile Nim game.

Problem 147, page 132

Problem 147 asks if an 8–marker unbalanced game could have more than 4 balancing moves. The 8-marker unbalanced game can be analyzed via the 4-pile Nim game. This game has at most three balancing moves (Why?) so the same is true of the 8-marker game.

Problem 148, page 133

To find a winning strategy you may wish to follow the approach of first comparing the strategies for Nim with Misère Nim for simple games. For example, proceed as follows:

(a) Determine the position that forces the next player to lose on the next move.

(b) Which positions, if any, are balanced when every pile has one stick?

(c) Which positions, if any, are balanced when exactly one pile has more than one stick?

(d) Which positions are balanced when more than one pile has more than one stick?

   Hint: Note the results of (b) and (c) and use the results of Nim.

(e) Describe a winning strategy for Misère Nim.

Try to use this outline to discover the strategy before checking the answer which now follows:

   Our winning strategy for Misère Nim follows this suggested outline.

(*a*) When there is only one stick left, the next player must take it and lose.

(*b*) When the number of piles is odd, the position is balanced. When that number is even, it is unbalanced. (Check this.) Note that this is the opposite of the situation in Nim.

(*c*) None! The position is unbalanced, since by taking sticks from the big pile, leaving either one or no sticks in that pile, depending on whether the number of piles is even or odd, a balanced position can be created. See (b).

(*d*) and (*e*). Suppose more than one pile has two or more sticks.

   1. If we can find a position such that any move that is made leads to a position that is unbalanced because of (b) or (c), that position is balanced.

   2. If not, we can try to postpone things to arrive at such a position eventually. We can achieve this by following the Nim strategy until we get to such a position!

Suppose we begin with a position that is balanced (in Nim). Any move our opponent makes creates an unbalanced position (in Nim). This cannot be a position with no piles with more than one stick (because the case we are considering assumes more than one such pile). If the opponent's move results in exactly one pile with more than one stick, we apply part (c).

If the position obtained still has more than one pile with two or more sticks, we rebalance as in Nim. Continuing this process we eventually arrive at a situation in which (b) or (c) applies.

Thus the Nim and Misère Nim games have exactly the same balanced positions except when case (b) applies.

Problem 149, page 133

Although Nim has a very subtle strategy that required us to learn the binary system and compute sums of binary digits, the strategy for Reverse Nim is quite easy.

We observe quickly that the balanced positions are those with an even number of sticks in each pile. The final position of no sticks satisfies the condition since zero is an even number. Any move from such a position leaves at least one pile with an odd number of sticks. And by taking one stick from each pile with an odd number of sticks, we have restored a position with all piles having an even number.

Problem 150, page 133

Don't jump to the conclusion that the balanced positions are those with all piles odd. Verify that that won't work

Does the term "reverse" in the title of this section suggest anything? If you think about that, you might see the answer. If not, see Section 3.8.1.

Problem 151, page 138

If there are n sticks in the pile (there is only one pile) then the only positions in the game are $x = 0, 1, 2, \ldots, n$. Certainly $g(0) = 0$ and $g(1) = 1$. You can use induction to prove that $g(x) = x$.

Problem 152, page 139

Let us use $g(a, b, c)$ to denote the value of the Sprague-Grundy function for a position (a, b, c) in the game. There are 24 possible positions in all in this particular game and we need to compute $g(a, b, c)$ for each.

Certainly $g(0, 0, 0) = 0$. These positions all lead directly and only to $(0, 0, 0)$:

$$(0, 0, 1), (0, 1, 0), \ (1, 0, 0)$$

and so each of these must be assigned a value of 1. The position $(0, 0, 2)$ leads only to

$$(0, 0, 0) \text{ and } (0, 0, 1)$$

and so $g(0, 0, 2) = 2$ (must be different from 0 and 1). The position $(0, 0, 3)$ leads only to

$$(0, 0, 0), (0, 0, 1) \text{ and } (0, 0, 2)$$

and so $g(0, 0, 3) = 3$ (must be different from 0 and 1 and 2).

The position $(1, 1, 0)$ (which we happen to know is balanced) leads only to $(1, 0, 0)$ and $(0, 1, 0)$ both of which have a g value of 1. Thus $g(1, 1, 0) = 0$.

Continue in this way working from the end of the game backwards. Note that we cannot determine $g(1, 2, 3)$ until we know all Sprague-Grundy values of all the positions

$$(0, 2, 3), (1, 1, 3), (1, 0, 3), (1, 2, 2), (1, 2, 1), \text{ and } (1, 2, 0).$$

(We don't yet.) Then we would pick for $g(1, 2, 3)$ the smallest nonnegative number that hasn't been assigned for these positions.

While we may lose patience with this procedure it is ideally suited to computer programming. Thus, in practice, computing the Sprague-Grundy function for all positions in a reasonably sized game takes no time at all.

Problem 153, page 139

First begin by computing the Sprague-Grundy function for a number of positions in the game. Start at the lowest depths and work up. This is rather tedious but will lead to an understanding of how this works. Figure 3.42 shows Depth and Sprague-Grundy numbers for various positions in 2–pile Nim. Depth is defined in Section 3.2.6.

Position (m,n)	Depth	$m \oplus n$	Position (m,n)	Depth	$m \oplus n$
(0,0)	0	0	(2,4)	6	6
(1,0)	1	1	(5,1)	6	4
(0,1)	1	1	(1,5)	6	4
(1,1)	2	0	(6,0)	6	6
(2,0)	2	2	(0,6)	6	6
(0,2)	2	2	(4,3)	7	7
(2,1)	3	3	(3,4)	7	7
(1,2)	3	3	(5,2)	7	7
(3,0)	3	3	(2,5)	7	7
(2,2)	4	0	(6,1)	7	7
(3,1)	4	2	(1,6)	7	7
(1,3)	4	2	(7,0)	7	7
(3,2)	5	1	(0,7)	7	7
(4,1)	5	5	(4,4)	8	0
(1,4)	5	5	(3,5)	8	6
(5,0)	5	5	(5,3)	8	6
(0,5)	5	5	(6,2)	8	4
(3,3)	6	0	(2,6)	8	4
(4,2)	6	6	(7,1)	8	6

Figure 3.42: Sprague-Grundy numbers for 2–pile Nim.

The table shows our computations for the Sprague-Grundy numbers up to a few at depth 8. Let us illustrate the method by showing that

$$2 \oplus 3 = 1.$$

The Grundy number for $(2,3)$ is not completely easy to compute, but it is a straightforward computation. Just look at all the positions next after $(2,3)$:

$$(2,3) \rightsquigarrow (2,2), (2,1), (2,0), (1,3), (0,3)$$

and the five Grundy numbers for these positions are

$$0, 3, 2, 2, 3$$

as we have already computed since they are at lower depths in the game. The smallest number that does not appear is 1 so $\mathrm{Grundy}(2,3) = 1$ and consequently, as

$$\mathrm{Grundy}(p_1, p_2) = p_1 \oplus p_2$$

holds in our notation, then we can write $2 \oplus 3 = 1$.

Such a "sum" may at first appear to be rather mysterious perhaps, but not in binary:

$$2 \oplus 3 = \begin{pmatrix} 1 \\ 0 \end{pmatrix} \oplus \begin{pmatrix} 1 \\ 1 \end{pmatrix} = \begin{pmatrix} 0 \\ 1 \end{pmatrix} = 1.$$

Let us pick a few more mysterious sums from the table and display them in binary:

$$3 \oplus 5 = \begin{pmatrix} 0 \\ 1 \\ 1 \end{pmatrix} \oplus \begin{pmatrix} 1 \\ 0 \\ 1 \end{pmatrix} = \begin{pmatrix} 1 \\ 1 \\ 0 \end{pmatrix} = 6$$

and

$$6 \oplus 2 = \begin{pmatrix} 1 \\ 1 \\ 0 \end{pmatrix} \oplus \begin{pmatrix} 0 \\ 1 \\ 0 \end{pmatrix} = \begin{pmatrix} 1 \\ 0 \\ 0 \end{pmatrix} = 4.$$

Try a few more and you will doubtless see the pattern which the Nim-sum section which follows now explains. Try to verbalize what you have observed before reading on to a full description of what a Nim sum is.

Problem 154, page 139

The method we use is the same method that will work to prove the Sprague-Grundy theorem in Section 3.10.4. It is a good warm-up to that theorem to try to see how this works here.

At depth zero the statement is evidently true since it amounts only to the fact that the Grundy number for the end position $(0,0)$ in 2–pile Nim is exactly $0 \oplus 0 = 0$. Thus it is only the induction step that takes us some trouble.

Suppose that the position (p_1, p_2) is at a depth for which we know that, for all positions (m, n) at any lower depth, the Grundy number for (m, n) in Nim is exactly $m \oplus n$ where this is the Nim-sum (i.e., binary addition without carry). Our proof below uses the assumption that we already know this is true at any lower depth.

Let

$$b = p_1 \oplus p_2$$

In order for us to prove that $\text{Grundy}(p_1, p_2) = b$ we must show that both of these statements are true:

1. For every non-negative integer $a < b$, there is a follower of (p_1, p_2) in Nim that has Grundy value a.

2. No follower of (p_1, p_2) has the Grundy value b.

Then the Grundy value at (p_1, p_2), being the smallest value not assumed by one of its followers, must be b.

To show (1), let $d = a \oplus b$ and let k be the number of digits in the binary expansion of d, so that

$$2^{k-1} \leq d < 2^k$$

and d has a 1 bit in the kth position in the binary expansion.

We have to remember now that $d = a \oplus b$ and remember too how the binary without carry operation $\oplus$ works. Since $a < b$, b must have a 1 in the kth position and a must have a 0 there. Since

$$b = p_1 \oplus p_2$$

we see that p_1 [or perhaps p_2] would have to have the property that the binary expansion of p_1 [or perhaps p_2] has a 1 in the kth position.

Suppose for simplicity that it is the first case. Then

$$d \oplus p_1 < p_1.$$

Define

$$p_1' = d \oplus p_1.$$

The move from (p_1, p_2) to (p_1', p_2) is a legal move in 2–pile Nim and

$$p_1' \oplus p_2 = d \oplus p_1 \oplus p_2 = d \oplus b = (a \oplus b) \oplus b = a \oplus (b \oplus b) = a.$$

We have produced the move

$$(p_1, p_2) \rightsquigarrow (p_1', p_2)$$

for which

$$p_1' \oplus p_2 = a.$$

Since this position is at a lower depth we know (by our induction hypothesis) that

$$\text{Grundy}(p_1', p_2) = p_1' \oplus p_2 = a.$$

Thus the follower (p_1', p_2) in Nim has a Grundy number a. This verifies our first statement.

Finally, to show (2), suppose to the contrary that (p_1, p_2) has a follower with the same Grundy value b. We can suppose that this involves removing sticks from the first pile. (The argument would be similar if it involved the second pile.)

That is, we suppose that (p_1', p_2) is a follower of (p_1, p_2) and that

$$\text{Grundy}(p_1', p_2) = p_1' \oplus p_2 = p_1 \oplus p_2.$$

(Here we have again used our induction hypothesis since the position (p_1', p_2) is at a lower depth.) Just like in ordinary arithmetic (using $+$ instead of as here $\oplus$) we can cancel the two identical terms in the equation

$$p_1' \oplus p_2 = p_1 \oplus p_2.$$

and conclude that

$$p_1' = p_1.$$

But this is impossible since

$$p_1 > p_1'$$

since we have removed some sticks from the first pile. That completes the proof at the induction step and so the statement must be true at all depths.

Problem 156, page 143

We simply note that $m \oplus n = 0$ if and only if $m = n$.

Problem 157, page 143

Since $1 \oplus 2 \oplus 3 = (1 \oplus 2) \oplus 3 = 3 \oplus 3 = 0$ it follows that the position $(1, 2, 3)$ is balanced. The other computation,

$$2 \oplus 3 \oplus 4 = (2 \oplus 3) \oplus 4 = 1 \oplus 4 = 5 \neq 0,$$

shows that $(2, 3, 4)$ is not.

Problem 158, page 143

We have computed the Grundy number for this position to be

$$2 \oplus 3 \oplus 4 = (2 \oplus 3) \oplus 4 = 1 \oplus 4 = 5.$$

We know that $5 \oplus 5 = 0$ so

$$(5 \oplus 2) \oplus 3 \oplus 4 = 0$$

and

$$2 \oplus (3 \oplus 5) \oplus 4 = 0$$

and

$$2 \oplus 3 \oplus (4 \oplus 5) = 0.$$

Check each of these numbers in the table:

$$(5 \oplus 2) = 7 \text{ and } (3 \oplus 5) = 6 \text{ and } (4 \oplus 5) = 1.$$

The only one that helps is the last one which tells us to reduce the pile with 4 down to 1 to change this position to a balanced position. We could also increase the pile with 2 up to 7 or the pile with 3 up to 6 but the rules of Nim don't allow us to add sticks. (That would be playing the game backwards, returning to a previous balanced position.)

Problem 159, page 144

$13 \oplus 12 \oplus 8 = 9.$

Problem 160, page 144

Set up the problem this way:

$$
\begin{array}{rcccccccc}
38 & = & 1 & 0 & 0 & 1 & 1 & 0 \\
\oplus \ x & = & ? & ? & ? & ? & ? & ? \\
\hline
25 & = & & 1 & 1 & 0 & 0 & 1
\end{array}
$$

and remember to perform the binary addition without carry. Clearly x is 111111 in binary. (What's that in decimal notation?)

Problem 161, page 145

You can easily check that $n \oplus n = 0$ and so
$$n \oplus n \oplus n = (n \oplus n) \oplus n = 0 \oplus n = n.$$
In general, then, $n \oplus n \oplus n \oplus \cdots \oplus n$ is either 0 or n depending on whether you are summing an even or odd number of terms.

Problem 162, page 145

The associative rule
$$(m \oplus n) \oplus p = m \oplus (n \oplus p)$$
is stated in the lemma. The zero element of the group is 0 itself, since
$$(m \oplus 0) = 0 \oplus m = 0$$
and every element n has an inverse for the operation $\oplus$ since
$$n \oplus n = 0.$$
Thus this is a group, a commutative group in fact since $m \oplus n = n \oplus m$ is always true.

Problem 163, page 145

It is certainly true that
$$(3, 10, 12) \to (3 \oplus 5, 10, 12)$$
$$(3, 10, 12) \to (3, 10 \oplus 5, 12)$$
and
$$(3, 10, 12) \to (3, 10, 12 \oplus 5).$$
produce balanced positions but only one of these is a valid Nim move. We have to *subtract* sticks from one of the piles and two of these suggestions *add* sticks.

Problem 164, page 145

It is certainly true that

$$(3, 10, 12) \rightarrow (3, 10, [3 \oplus 10])$$
$$(3, 10, 12) \rightarrow (3, [3 \oplus 12], 12)$$

and

$$(3, 10, 12) \rightarrow ([10 \oplus 12], 10, 12)$$

produce balanced positions but only one of these is a valid Nim move. We have to *subtract* sticks from one of the piles and two of these suggestions *add* sticks.

Chapter 4

Links

Figure 4.1 shows three interlinked circles arranged in such a way that should any one of the three circles be cut and removed, the remaining two circles would become separated. This arrangement has been known for many centuries and, because of the number 3 and the special nature of the linking, has been used for various symbolic representations.

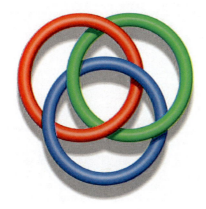

Figure 4.1: Borromean rings (three interlinked circles).

Some suggest that an image of God as three interlaced rings inspired Dante Alighieri (1265-1321). In his *Divina Commedia* he describes this vision:[1]

> Ne la profonda e chiara sussistenza
> de l'alto lume parvermi tre giri
> i tre colori e d'una contenenza;
> e l'un da l'altro come iri da iri
> parea reflesso, e'l terzo parea foco
> che quinci e quindi igualmente si spiri.
> —[Dante, *Paradiso*, §33, 115-120]

---

[1]Within the profound and shining subsistence of the lofty light appeared to me three circles of three colors and one magnitude; and one seemed reflected by the other, as rainbow by rainbow and the third seemed fire breathed forth equally from the one and the other.

On a more profane level the name that is most often
attached to these three interlocked circles arises from the
Borromeo family of 16th century Milan, who had such a
figure on their coat-of-arms. Many of our readers might
prefer to call these *Ballantine rings* since the three interlocked rings have appeared since 1879 as a company logo
for Ballantine Ale. The famous Ballantine three ring symbol (Purity, Body, Flavor) was, according to company folk-
lore, inspired by the wet rings left on a table as Peter Ballantine consumed his beer.

Figure 4.2: Bal-
lantine Ale

   In this chapter we consider a variety of problems related
to this construction. It is easy enough to design three circles
that interlink in the way the Borromean rings do. Could one do the same with
four or five circles? Or could we arrange for other kinds of linking properties,
say five rings linked together that do not fully separate unless two (any two) are
cut away?

   Our discovery process in this task is similar in many ways to the process that
we followed in our Tiling chapter. As before we don't see immediately how any
of the standard methods of arithmetic, algebra, or geometry could be brought
to bear on such problems. Once again we need to get a *feel* for the problem by
experimenting with a few examples.

4.1   Linking circles

Look at the two pairs of circles in Figure 4.3. Our sketch is meant to suggest
that they are curves in three dimensions.

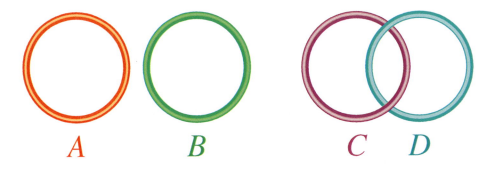

Figure 4.3: Four circles.

   The picture displays the fact that the circles *A* and *B* are not linked together
while the circles *C* and *D* are. This means that *C* and *D* cannot be separated
(without cutting or tearing). We are going to consider a certain class of problems
involving the ways in which three-dimensional curves can be linked. The curves
need not be circles.

4.1.1 Simple, closed curves

Before we state the first problem, we should make sure we agree on what a curve is or, more precisely, on what kinds of curves we shall be considering.

All of our curves are placed in *three dimensions* and all are *simple* and *closed* in the sense we now define.

Consider the five curves sketched in Figure 4.4.

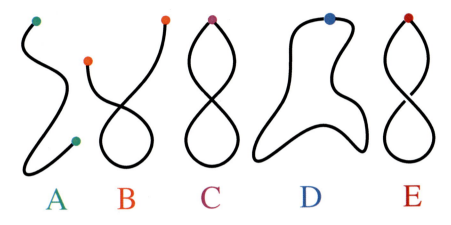

A   B   C   D   E

Figure 4.4: Simple curves, closed curves or not?

A curve is called *simple* if it does not cross itself. That means that in tracing out the curve (starting at any point) no point except, possibly, the beginning and end of the tracing is encountered more than once. It is called *closed* if it "ends where it starts."

Thus *A* is simple but not closed, *B* is neither simple nor closed, *C* is closed but not simple and *D* and *E* are simple and closed. Curve *C* is depicted in two dimensions and has a crossing point. Curve *E* is intended to be three-dimensional. Part of the curve—the part that appears as a break—lies below the darker part that "appears to cross" the broken part. Curve *E* does *not* cross itself.

The curves that are suitable for the discussion that follows must be both simple and closed. For that reason, we shall always assume (in this chapter) that when we use the term *curve* we mean "simple, closed curve." All of our curves are given in three-dimensions.

4.1.2 Shoelace model

You might find it desirable to make a model with which to experiment. Such models will be of use throughout this chapter, so it is a good idea to make such a model now. This can be done in a variety of ways. For example, using the outer edges of paper plates or wire, construct several rings to represent the curves. Two or three of these rings should be pre-cut in such a way that they can be easily removed from a configuration without disturbing the rest of the configuration. We refer to this as "cutting away a curve." You will also need something more flexible for your experimentation. A long shoelace or ribbon will do. Figure 4.5 shows[2] some equipment that might be used.

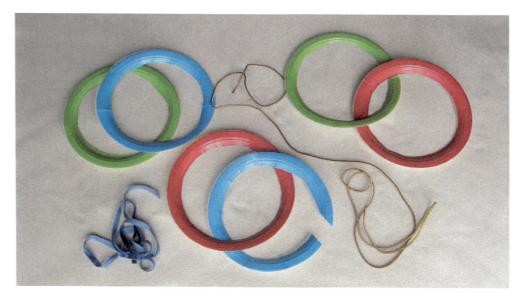

Figure 4.5: Equipment for making models.

4.1.3   Linking three curves

Do you think it is possible to construct three curves that are linked in such a way that, if we cut away one of the three curves, the remaining two will remain linked? By this we mean that *no matter which of the three curves* we cut away, the remaining two *cannot* be separated without cutting or tearing. (Pulling is all right.)

Here is a different but related question:

> Is it possible for three curves to be linked together in such a way that no curve can be separated from the configuration without cutting or tearing, but if one is cut away, the remaining two can be separated?

As before, we mean by this that *no matter which of the three* is cut away, the remaining two *can* be separated. Thus, in a sense, the "break point" is at two curves: the configuration "hangs together" as originally constructed, but removal of a single curve causes the remaining configuration to "fall apart."

The Borromean rings of Figure 4.1 give a positive answer to the latter question. Check that Figure 4.1 does answer the second question but not the first.

---

[2]Photo courtesy of Curry Sawyer.

Problem 165 *Is it possible to construct three curves that are linked in such a way that if we cut away any one of the three curves, the remaining two will remain linked?* Answer □

Problem 166 *Without looking again at Figure 4.1, describe three curves linked together in such a way that no curve can be separated from the configuration without cutting or tearing but, if one is cut away, the remaining two can be separated?* Answer □

4.1.4   3–1 and 3–2 configurations

Let us call the configuration that we constructed in Problem 165, a 3–1 configuration. The "3" refers to the fact that there are three curves; the "1" indicates that the *breaking point* is at 1—the configuration can't be separated without cutting or tearing until we get down to one curve.

We call the second configuration (the Borromean rings) a 3–2 configuration because there are three curves linked in such a way that the configuration hangs together, but removal of any one of the three curves by cutting causes the other two curves to fall apart with no more than a *pull* in the appropriate place. The breaking point is at 2.

4.1.5   A 4–3 configuration

What should we mean by a 4–3 configuration? Well, that would be a configuration of four curves linked together in such a way that they hang together but cutting away one of the curves causes the remaining three to fall apart. And this is true no matter which of the four curves is cut away. In short the breaking point is at three curves—any three.

Problem 167 *Do you think it is possible to construct a 4–3 configuration?*
Answer □

4.1.6   Not so easy?

Experience has shown that many students have difficulty in their first attempts to construct a 4–3 configuration. As before, we may as well begin by placing three separated curves near each other. This represents the initial setup. We then try to weave a fourth curve (the shoelace) through them in order to construct the 4–3 configuration. Then, no matter how this is done, we can at least be assured that removal of the shoelace will cause the remaining three curves to fall apart. (They are already separated.)

But then the trouble begins. Unless we have a very usable model, or three or four hands, or a friend to hold part of the model while we do our weaving, we wind up with knots in the shoelace, the model falling on the floor or other such problems.

And to make things worse, when we finally get things together it almost works. But then we try to cut away a curve to "check it out" and see it does not quite work. A little change might do it, but by now we forgot what we did to make it almost work. Frustrating. We know. We tried and it happened to us.

There must be a better way. Even if we had a good model, four hands, a friend to help, and we solved this problem what will we do when we get to more complicated linking problems?

4.1.7 Finding the right notation

Let's return to the 3–2 configuration. We might observe that our construction method solved the problem. We could make a model that worked, but the language was awkward and communication was difficult. We had to talk about the "curve on the right," and the "remaining curve." And we drew a picture which suggested words such as the following:

> "go through the curve on the left from top to bottom, then the curve on the right from top to bottom, then the curve on the left from bottom to top, then the curve on the right from bottom to top and then return to the starting point."

(We might add "Do not pass GO and do not collect \$200.")

We could simplify things considerably with a bit of *notation*. If we label the curve on the left A and the curve on the right B, our description could be written ABA^bB^b.

Simple, isn't it? After working a bit with the notation you will find that it has made things quite easy. The notation contains all essential ingredients. The expression

$$ABA^bB^b$$

is read from left to right and translates into

> "Go through A, then B, then A backwards, then B backwards."

(The fact that we ended where we started is understood and the parenthetical reference to the game of *Monopoly*™ is of course unnecessary.)

All we needed was a labeling of the curves (other than the shoelace) and our notation provides a "recipe" or a set of directions for the construction. There are, of course, two things we are assuming tacitly: we are assuming the first curves have already been labeled and placed appropriately, and we are assuming that in passing through a given curve there are two directions we could use, one positive and one negative. Thus the notation A and A^b represent passing through A, but in the first case the shoelace passes through A in the direction we called positive and on the second it was in the opposite direction.

Which direction we designate as positive (for each curve) is immaterial, but once we have chosen it, it is essential that we remain consistent. We shall answer Problem 167 soon. If you were unable to solve it, try it again, but first do the problems below.

Pulling is allowed, but remember, the weaving must *end* where it started and the shoelace is assumed solidly attached at its "two ends." (The quotation marks are there because the shoelace is just a physical model of the concept of simple closed curve and we are not really thinking of such a curve as having *ends* any more than we do of a rubber band or a circle.)

Problem 168 *Let A and B be separated curves. Weave your shoelace through the curves A and B according to each of the descriptions below:*

$$BAB^bA^b, \; A^bBAB^b, \; AA^bBB^b, \; AAA^bA^b, \; and \; ABB^bA^b.$$

Provide a pencil sketch of the result (if you can) as well as following the instructions on your physical model. (This problem and the next two should be done together. Use your model.) Answer □

Problem 169 *Which of the descriptions in Problem 168 lead to a 3–2 configuration?*

Answer □

Problem 170 *Which of the descriptions in Problem 168 lead to a configuration that can be separated without cutting or tearing.* Answer □

Problem 171 *Study the results of Problem 168, Problem 169, and Problem 170. What patterns do you notice in those which give rise to 3–2 configurations? Do you notice any symmetries. Do you notice any* delayed undoings *of things already done? Try to articulate for yourself what makes the 3–2 configuration work.* Answer □

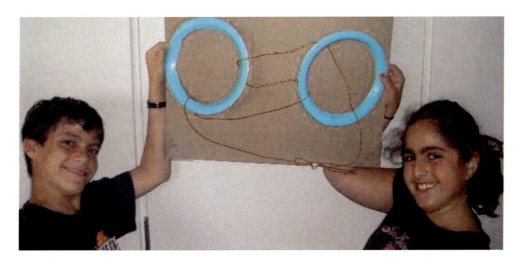

Figure 4.6: Cole and Eva with model.

Problem 172 *Figure 4.6 shows Cole and Eva with a model of a configuration.*
Is this one of the configurations we have discussed?   Answer ☐

Problem 173 *Construct a 4–3 configuration.*   Answer ☐

4.2 Algebraic systems

Perhaps you have noticed similarities between our notation and things you re-member from arithmetic or algebra. First of all, an expression such as *AB* is reminiscent of the operation of multiplication. Of course, in our setting *A* and *B* do not represent numbers. Far from it. And writing *B* next to *A* doesn't mean we *multiply A* by *B*. That wouldn't make sense in our setting.

To get a better idea of what we mean when we say that our notation is reminiscent of algebra, we shall undertake a rather long-winded digression.

An *algebraic system* consists of a collection of objects (e.g., numbers), one or more operations (e.g., addition or multiplication), and some rules or axioms governing the ways in which these objects can be combined. Familiar to us from the rules of arithmetic are the commutative and associative laws for mul-tiplication: whatever numbers *A*, *B* and *C* we choose, it will always be true that $AB = BA$ and that $(AB)C = A(BC)$. For example $2 \times 3 = 3 \times 2$ since both equal 6 and

$$(2 \times 3) \times 4 = 2 \times (3 \times 4)$$

since both equal 24. This is a theorem: technically 2×3 and 3×2 mean two different things, but they can be proved to be equal. Similarly $(2 \times 3) \times 4$ and $2 \times (3 \times 4)$ mean two different things but they can be proved to be equal for numbers and the operation of multiplication. It is rules similar to these that we are interested in when discussing algebraic systems. Indeed, in some algebraic systems, things look pretty much the same as they do in ordinary algebra or arithmetic.

4.2.1 Some familiar algebraic systems

Here are some examples of algebraic systems with which you may or may not be familiar.

1. The objects are *polynomials*. They can be added or multiplied and the laws of combination that apply to numbers apply here as well. For exam-ple

$$[x^2 + x + 1] + [3x - 2] = x^2 + 4x - 1$$

and

$$[x^2 + x + 1] \times [3x - 2] = 3x^3 + x^2 + x - 2.$$

2. The objects are *functions*. They, too, can be added or multiplied and the laws apply. For example $x^2 + \sin x$ is obtained by adding the function x^2 to the function $\sin x$. Furthermore, $x^2 + \sin x = \sin x + x^2$.

3. The objects are *matrices* of fixed dimension. Addition and multiplication are defined but the commutative law of multiplication does not hold. For example

$$\begin{pmatrix} 1 & 0 & 0 \\ 1 & 2 & 3 \\ -1 & 0 & 5 \end{pmatrix} + \begin{pmatrix} 0 & 1 & 1 \\ 2 & 0 & -1 \\ -2 & 0 & -1 \end{pmatrix} = \begin{pmatrix} 1 & 1 & 1 \\ 3 & 2 & 2 \\ -3 & 0 & 4 \end{pmatrix}.$$

4. The objects are *vectors*. Again addition can be defined in a natural way and the usual laws hold with respect to the operation of addition. For example

$$(1,2,3) + (5,6,-3) = (6,8,0).$$

4.2.2 Linking and algebraic systems

In the setting of our linking problems, each configuration gives rise to an algebraic system. The configuration is the *starting set-up*, for example, two separated curves A and B. Each *shoelace* gives rise to an *object* of this algebraic system via its *formula*. The formula is just a string of the letters (in this case A and B) that corresponds to the way the shoelace links the curves in the starting set-up. We discuss this in more detail later in this chapter in Section 4.9.2.

For example, the expression

$$ABA^b BB$$

would represent the curve that goes through A, then B, then A backwards then twice through B and then returns to the starting point. The *operation* for the system can be described as *one object of the system following another*. As an example, if $X = AB$ and $Y = A^b BB$, then $XY = ABA^b BB$.

In order to create an algebraic system we must do several more things: we must decide what it means for two of our objects to be *equal* and we must determine what the basic *laws of combination* are.

4.2.3 When are two objects equal?

What should it mean for two objects to be *equal?* This does not mean that they are the exact same expression. It depends on the algebraic system to interpret equality. For example in the elementary theory of fractions $1/2$ and $2/4$ and $17/34$ are defined to be equal, even though they are not identical expressions.

Well, for our purposes in the linking problems, it would be natural to define equality in such a way that two objects are equal if and only if they have exactly the same linking properties. For example,

$$AA^b, \ BB^b, \ A^bA, \ B^bB, \ AB^bBA^b, \ \text{and} \ AAAA^bA^bA^b$$

all are different expressions for the shoelace which links neither A nor B. That is, an appropriate pull on the shoelace will set it free from the curves A and B. Similarly,

$$A, \ ABB^b, \ AA^bA, \ \text{and} \ BAA^bB^bA$$

all represent the same linking properties: a curve that, in effect, goes through A and returns home.

Problem 174 *Interpret, for this problem, A and B as positive numbers and interpret* A^b *and* B^b *to be the reciprocals* $1/A$ *and* $1/B$. *Compute each of the expressions*

$$AA^b, \ BB^b, \ A^bA, \ B^bB, \ AB^bBA^b, \ \text{and} \ AAAA^bA^bA^b.$$

Answer □

Problem 175 *Under the same interpretation as in Problem 174 compute each of the expressions*

$$A, \ ABB^b, \ AA^bA, \ \text{and} \ BAA^bB^bA.$$

Answer □

4.2.4   Inverse notation

For numbers we wouldn't write A^b for the reciprocal of A. We would write A^{-1}. Thus $AA^{-1} = A^{-1}A = 1$ for numbers; e.g., $5 \times 5^{-1} = 5^{-1} \times 5 = 1$. This suggests that we should use the same notation for our linking instruction to go backwards. This suggests, too, that we should use the symbol 1, not for the number one, but for any curve that doesn't link either A or B. Our notation becomes even more reminiscent of algebra.

We would see quickly, for example, that in our setting

$$AA^{-1}, \ ABB^{-1}A^{-1}, \ \text{or} \ A^{-1}AAA^{-1}$$

are all equal to 1. It would also be true, just as in ordinary arithmetic, that

$$1A = A1 = A.$$

(If a curve doesn't link anything and then links A, in effect it has linked only A).

Thus we shall decide on the more natural algebraic notation A^{-1} to represent the instruction to go through A backwards. We can then use our elementary skills in algebra to help us work out the effect of such complicated expressions as we saw in Problem 174 and Problem 175. Those expressions now assume the simpler and more familiar form

$$AA^{-1}, \ BB^{-1}, \ A^{-1}A, \ B^{-1}B, \ AB^{-1}BA^{-1}, \ \text{and} \ AAAA^{-1}A^{-1}A^{-1}$$

all of which reduce easily to 1 and
$$A, \; ABB^{-1}, \; AA^{-1}A, \;\; \text{and} \; BAA^{-1}B^{-1}A$$
all of which reduce easily to A.

4.2.5 The laws of combination

We can now easily verify that

- The commutative law *fails*: AB and BA are, in general, different.

- The associative law is valid: $(A(BC) = (AB)C)$ is always true.

You can use your models to check these facts, or you can just give simple arguments to verify this. For example, since
$$ABA^{-1}B^{-1}$$
gives rise to the 3–2 configuration while
$$AA^{-1}BB^{-1} = 1$$
we see the commutative law fails. On the other hand, $A(BC)$ represents going through A, then through B and C, then home. That has exactly the same effect as the linking $(AB)C$.

Problem 176 *Use your model to verify that*
$$ABA^{-1}B \neq 1.$$
(The commutative law does not hold.) $\square$

4.2.6 Applying our algebra to linking problems

Where does all this get us? For one thing, it allows us in some cases to check the linking effect algebraically without reference to a picture or model. For example, we can reduce the expression
$$ABA^{-1}BB^{-1}AA^{-1}B$$
algebraically as follows
$$ABA^{-1}BB^{-1}AA^{-1}B = ABA^{-1}11B = ABA^{-1}B$$

Thus, in effect, our shoelace has gone through A, then B, then A backwards, then B. We can reduce this no further.

Cutting away But there is something even more useful contained in our algebraic system. Our algebraic system is particularly useful when it comes to *cutting away* one of the circles. Consider the expression
$$ABA^{-1}B^{-1},$$

which we saw gave rise to the 3–2 configuration when applied to two separated circles. What happens when we cut away the circle A? We saw that the shoelace and B were not linked.

But we can tell this *just by inspecting* the expression

$$ABA^{-1}B^{-1}.$$

How is cutting away A reflected in the expression $ABA^{-1}B^{-1}$? If A is cut away, then we in effect have a new *starting set up*, the one circle, B. Where the shoelace originally went through A is now an empty space.

Simply remove the symbol A (and A^{-1}) whenever it appears in the expression $ABA^{-1}B^{-1}$. We arrive at the expression BB^{-1} which of course equals 1, a curve that links nothing.

What we have just seen is an essential step in our attempt to construct configurations exhibiting certain linking properties. To make sure we understand it, use your model to answer the next problem where one circle has been cut away.

Problem 177 *Start with three separated circles A, B and C. What happens when B is cut away from the expressions ABC, $ABCA^{-1}B^{-1}C^{-1}$, and $ABA^{-1}B^{-1}$.*

Answer □

4.3   Return to the 4–3 configuration

Now we return to the 4–3 configuration. We begin with three separated circles A, B and C. We wish to wind our shoelace through A, B and C in such a way that the configuration hangs together, but removal of one of the curves causes the remaining three curves to fall apart. This must be true no matter which of the curves we remove.

Translated into our algebraic setting, we seek an expression involving the letters A, B, and C that does not reduce to an expression with fewer letters, but removal of a single letter causes the expression to reduce to 1.

Before trying to achieve this, observe that for the 3–2 configuration, the expression $ABA^{-1}B^{-1}$ does not reduce, but indeed removal of either A or B causes the resulting expression to collapse to 1.

4.3.1   Solving the 4–3 configuration

There are a number of ways of achieving the 4–3 configuration. Here is one of them, which you may have discovered

$$ABA^{-1}B^{-1}CBAB^{-1}A^{-1}C^{-1}$$

Try it on your model, making sure to avoid *knots*. Observe, removal of A results in the expression

$$BB^{-1}CBB^{-1}C^{-1} = 1C1C^{-1} = CC^{-1} = 1.$$

The same is true if B is removed. If C is removed, we obtain

$$ABA^{-1}B^{-1}BAB^{-1}A^{-1} = ABA^{-1}1AB^{-1}A^{-1}$$

$$= AB1B^{-1}A^{-1} = ABB^{-1}A^{-1} = A1A^{-1} = AA^{-1} = 1.$$

Undoing   If we understand the algebraic model, this last computation can be greatly simplified. We must observe only that removal of C causes successive *collapses from inside to outside.* Look at the expression for the 4–3 configuration carefully to see what's involved. Each action is *undone* a little later. For A and B, the undoing is postponed only one step, but for C it is postponed several steps. Let's see why that works. Hidden in the expression is another *undoing.* The entire expression

$$ABA^{-1}B^{-1}$$

is undone by the expression

$$BAB^{-1}A^{-1}$$

because

$$\left(ABA^{-1}B^{-1}\right)\left(BAB^{-1}A^{-1}\right) = 1.$$

In our algebraic notation this means that

$$\left(ABA^{-1}B^{-1}\right)^{-1} = \left(BAB^{-1}A^{-1}\right).$$

Observe that to achieve this, we have undone each link in $ABA^{-1}B^{-1}$ but in the reverse order. It's like putting on your socks and shoes. To undo that action, you undo each step but in the reverse order, you first take off your shoes and then your socks. Or, at least, we do. The same is true for any linking. To undo it, that is to find an expression for the inverse, you undo each link but in the reverse order.

Problem 178 *Check each of these statements on your model:*

$$\left(ABCB^{-1}\right)^{-1} = BC^{-1}B^{-1}A^{-1} \ \ and \ \ \left(AB^{-1}\right)^{-1} = BA^{-1}.$$

□

Problem 179 *Compare the expressions for the shoelace in the 3–2 and 4–3 configurations:*

- *The 3–2:* $ABA^{-1}B^{-1}$.

- *The 4–3:* $ABA^{-1}B^{-1}CBAB^{-1}A^{-1}C^{-1} = (ABA^{-1}B^{-1})C(ABA^{-1}B^{-1})^{-1}C^{-1}$.

Answer □

Problem 180 *Construct a 5–4 configuration.* Answer □

4.4   Constructing a 5–4 configuration

Once we understand what makes the construction of the 4–3 configuration work, we find the problem of constructing a 5–4 configuration a bit less challenging. We begin with four separated circles and label them A, B, C, and D. We now wish to weave the fifth curve through these circles in an appropriate way. We know by now that we must undo each action at a later time. Just how much later this should be might now be apparent.

4.4.1   The plan

In case it is not yet entirely clear, consider the following plan:

- For a 3–2 configuration, begin with two separated circles A and B and this expression for the last curve

$$ABA^{-1}B^{-1}.$$

- For a 4–3 configuration, begin with three separated circles A, B, and C and this expression for the last curve
$$ABA^{-1}B^{-1}CBAB^{-1}A^{-1}C^{-1}.$$

Noting, as we did before, that the expression for the fourth curve in the 4–3 configuration can be written in the form $XCX^{-1}C^{-1}$, where
$$X = ABA^{-1}B^{-1}$$
we are naturally led to try the formula
$$YDY^{-1}D^{-1}$$
for the fifth curve of the 5–4 configuration, with
$$Y = ABA^{-1}B^{-1}CBAB^{-1}A^{-1}C^{-1}$$
This becomes
$$ABA^{-1}B^{-1}CBAB^{-1}A^{-1}C^{-1}DCABA^{-1}B^{-1}C^{-1}BAB^{-1}A^{-1}D^{-1}.$$

4.4.2   Verification

To verify that our proposed solution
$$ABA^{-1}B^{-1}CBAB^{-1}A^{-1}C^{-1}DCABA^{-1}B^{-1}C^{-1}BAB^{-1}A^{-1}D^{-1} \qquad (4.1)$$
works we must check these two things:

- The entire configuration hangs together.

- The removal of any one of the five curves causes the remaining curves to fall apart.

That the configuration hangs together is probably clear by now. You can check it with a model, but you can see it more easily from the algebraic model. No reduction is possible in the algebraic expression (4.1). This is because the only admissible simplifications allowable are to replace an expression such as XX^{-1} by 1 and to then *drop* the 1, and no such expression appears in (4.1).

To check that removal of any one of the curves causes the remaining curves to fall apart, we must verify that removal of a single letter whenever it appears causes the entire expression to reduce to 1. We do this for the letter A and leave it to you to verify that this happens when the letters B or C or D are removed instead of A. Removal of A leads to the expression

$$BB^{-1}CBB^{-1}C^{-1}DCBB^{-1}C^{-1}BB^{-1}D^{-1}$$

$$= 1C1C^{-1}DC1C^{-1}1D^{-1} = CC^{-1}DCC^{-1}D^{-1}$$

$$= 1D1D^{-1} = DD^{-1} = 1.$$

Do you think you could have constructed the 5–4 configuration using only trial and error on your model? Note that the *shoelace* had to go through the curves 22 times in all. Or perhaps you found a simpler solution.

4.4.3 How about a 6–5 configuration?

Suppose we now wanted to construct a 6–5 configuration. Does our solution of the 5–4 configuration lead us on to more complicated problems?

The pattern is probably clear by now, but the notation is getting rather out of hand. Note, for example, that the

- 3–2 configuration required 4 winds by the shoelace.

- the 4–3 configuration required 10 winds by the shoelace.

- the 5–4 configuration required 22 winds by the shoelace.

A bit of reflection would show that the 6–5 configuration would require 46 winds by the shoelace. Some simplification of notation is necessary, or at least desirable, here.

We note that each of the three configurations we have constructed so far is of the form

$$UVU^{-1}V^{-1}$$

where U is an expression involving, perhaps, several winds and V represents a single wind. For example, in the 4–3 configuration,

$$U = ABA^{-1}B$$

and

$$V = C.$$

This suggests our new notation.

4.4.4   Improving our notation again

Let us introduce some short-hand notation. If U and V are any expressions involving several letters (such as A, B, C, D, E, etc.), let us write (U,V) to represent the expression

$$(U,V) = UVU^{-1}V^{-1}.$$

Thus for $U = ABA^{-1}B^{-1}$ and $V = C$ the expression (U,V) becomes

$$(U,V) = \left(ABA^{-1}B^{-1}\right)C\left(ABA^{-1}B^{-1}\right)^{-1}C^{-1}$$
$$= ABA^{-1}B^{-1}CBAB^{-1}A^{-1}C^{-1}.$$

This gives us a compact notation.

Problem 181 *Verify that, in this notation, the fifth curve for the 5–4 configuration becomes*

$$(((A,B),C),D).$$

□

Problem 182 *Write the sixth curve of the 6–5 configuration in the compact form.*   Answer □

Problem 183 *Show that if $A = 1$ in the expression $((A,B),C)$, then*

$$((A,B),C) = 1.$$

Answer □

Problem 184 *Show that if $B = 1$ in the expression $((A,B),C)$, then*

$$((A,B),C) = 1.$$

□

Problem 185 *Show that if $C = 1$ in the expression $((A,B),C)$, then*

$$((A,B),C) = 1.$$

□

4.5   Commutators

The expression (A,B) has a name in the study of algebra. It is called the *commutator* of A and B. More generally, if X and Y are any expressions in several letters, then (X,Y) is called the *commutator* of X and Y. For instance, if

$$X = ABA^{-1}B^{-1}$$

and $Y = C$ then

$$(X,Y) = XYX^{-1}Y^{-1} = ABA^{-1}B^{-1}CBAB^{-1}A^{-1}C^{-1}$$

which we saw gives the fourth curve in the 4–3 configuration. Since

$$ABA^{-1}B^{-1} = (A,B),$$

the expression (X,Y) above equals $((A,B),C)$. This is a commutator, one of whose terms is itself a commutator. We call such an expression a *compound commutator*. Because of what will follow shortly, it is important to understand this commutator notation. As practice with the notation, do each of the following computations before proceeding further.

Problem 186 *Show that*

1. $((A,B),A^{-1}) = ABA^{-1}B^{-1}A^{-1}BAB^{-1}$

2. $((A,A^{-1}),B) = 1.$

3. $(((A,1),C),D) = 1.$

Answer □

Problem 187 *To see the importance of putting in all commas and parentheses in the commutator notation, verify that in general*

1. $(AB) \neq (A,B).$

2. $(AB)^{-1} \neq (A,B)^{-1}.$

3. $(AB,C) \neq (A,B)C \neq ((A,B),C).$

□

Problem 188 *One reason that (X,Y) is called the commutator of X and Y is that $XY = YX$ if and only if $(X,Y) = 1$. Prove this.* Answer □

4.6 Moving on.

So far, we have seen how to construct the 3–2, 4–3, 5–4 and 6–5 configurations. All of these are configurations of the type $(n+1)$–n; i.e., the breaking point occurs when a single curve is removed from the configuration.

What if we want the breaking point to occur somewhat later? For example, how could we construct a 4–2 configuration? First, we must be sure we understand what a 4–2 configuration is. It consists of four curves linked together in such a way that the entire configuration hangs together, that removal of a single curve causes the remaining three to hang together, but removal of a second curve causes the remaining two to fall apart.

Before embarking on the construction of a 4–2 configuration, let us pause for a moment to take stock of where we are.

4.6.1 Where we are.

After some trial and error with our models we discovered how to construct the 3–2 configuration. Perhaps we were also successful in constructing the 4–3 configuration by this method. Perhaps not. In any case things quickly became too complex to rely on trial and error and on our simple model. We arrived very naturally at an algebraic formulation of our problems.

It amounted to beginning with an appropriate placement of our first few curves and then writing down an expression for the last curve. This expression had to link all the existing curves and had to have the property that removal of a single letter caused the entire expression to reduce to 1. After a while, we saw the advantage of compact notation, and we introduced the idea of a commutator. All this allowed us to see the structure of the configurations in an algebraic setting. There were several new concepts, all of which evolved naturally:

- Algebraic expressions for the last curve.

- How to simplify such algebraic expressions, using ideas suggested from elementary algebra.

- The idea of a commutator and the natural extension to a compound commutator.

By successive compoundings of the commutator, we were able to construct more complicated configurations.

We need only one more idea to show us the way to constructing any configuration we wish. This idea will arise in connection with the 4–2 configuration. It will become clear a bit later.

4.6.2 Constructing a 4–2 configuration.

Let's get started with the construction of a 4–2 configuration. First of all, what do we start with? Since the breaking point is "2," we may as well begin with two separated circles A and B. This way, when the two new curves that we shall add are removed, we will end up with the two separated curves A and B that we started with.

Now what? That is, how should the third curve be woven through A and B?

Before attempting Problem 189 and Problem 190 try to determine what concepts are involved. Whether or not you obtain solutions, check our answer. Much of the reasoning in those answers will be needed in constructing the configurations that follow the 4–2 configuration.

Problem 189 *What expression should represent the third curve C?*

Answer □

Problem 190 *What expression should represent the fourth curve, D, in the 4–2 configuration?* Answer ☐

Problem 191 *Compare the two solutions*

$$ABCA^{-1}B^{-1}C^{-1}$$

and

$$(A,B)(A,C)(B,C).$$

How many winds does each require? Would either of these methods be useful for obtaining other configurations such as the 5–2 or 5–3 configuration?

Answer ☐

4.6.3 Constructing 5–2 and 6–2 configurations.

Let us try to imitate the two methods we used for the 4–2 configuration to construct a 5–2 configuration.

Problem 192 *Construct a 5–2 configuration. Begin with two separated curves A and B and determine formulae for the remaining curves C, D and E.*

Answer ☐

Problem 193 *Construct a 6–2 configuration. Use the method that starts off with ABCD ....* Answer ☐

Problem 194 *Construct a 6–2 configuration. Use the method that starts off with $(A,B)(A,C)(A,D)$ ....* Answer ☐

4.7 Some more constructions.

What about the 5–3 configuration? How can we use what we have already learned? How do we begin? The last question is, of course, easy to answer. Since the break point is "3," we must *begin* with three separated curves: *A*, *B* and *C*.

Arguing as before, we want our fourth curve to wind through these three curves in such a way as to form a 4–3 configuration.

Problem 195 *Now what?* Answer ☐

4.8 The general construction

We are now ready to understand the general construction. Before proceeding to that, construct each of the configurations below. Do this by indicating how many separated curves start the process and then giving the expression for the remaining curves, stating what the result is after each curve is added.

For example, the following format for the 6–4 configuration can serve as a model. Begin with four separated circles, A, B, C, and D.

Add	Formula	Config-uration
E	$(((A,B),C),D)$	5–4
F	$(((A,B),C),D)(((A,B),C),E)(((A,B),D),E)(((B,C),D),E)$	6–4

Problem 196 *Construct a 6–3 configuration. Verify that your constructions works.* □

Problem 197 *Construct a 7–3 configuration. Verify that your constructions works.* □

Problem 198 *Construct an 8–4 configuration. Verify that your constructions works.* Answer □

4.8.1   Introducing a subscript notation

The configurations that Problems 196–198 ask for should offer no serious difficulty (except that they take increasing amounts of space to write down).

Consider Problem 198. It asked for eight curves linked together in such a way that the break point occurs at 4 curves. The eighth curve, H, was added to a 7–4 configuration. The formula for the eighth curve involves 35 compound commutators on four letters, for example $(((A,B),C),D)$. Each such commutator has twenty-two winds. The eighth curve thus had $35 \times 22 = 770$ winds. Even the short-hand commutator notation would involve writing down 35 compound commutators. Note that replacing A, B, C or D with a 1 causes all commutators involving that letter to collapse, leaving a 7–4 configuration as required.

More notation   The time has come, once again, to introduce some further short-hand notation. Let's discuss this to see what kind of notation might be useful.

First, let's look ahead. The 8–4 configuration would be a nuisance to write down in full commutator notation—but we could still do it. What if, for example, we instead wanted to determine the 30th curve in a 30–20 configuration? We see that our alphabet, with only 26 letters, is inadequate.

Of course, we could add the Greek ($\alpha, \beta, \gamma, \dots$), Hebrew (, ℶ, ℸ, ℷ…), and Old-German ($\mathfrak{A}, \mathfrak{B}, \mathfrak{C}, \dots$) alphabets in order to obtain more symbols to use. But then what would we do if we wanted to construct a 300–200 configuration? Or a 3,000–2000 configuration?

There must be a better way! There is. It is, in fact, quite simple (though it may appear complicated at first). First we can solve our dilemma of running out of letters of the alphabet by introducing subscripts. Thus, instead of writing the fourth curve in a 4–3 configuration as

$$((A,B),C),$$

we call our first three curves A_1, A_2 and A_3 instead of A, B and C. The fourth curve A_4 then has the formula

$$A_4 = ((A_1,A_2),A_3).$$

Then, for example, the fifth curve of the 5–3 configuration will have the formula

$$A_5 = ((A_1,A_2),A_3)((A_1,A_2),A_4)((A_1,A_3),A_4)((A_2,A_3),A_4).$$

4.8.2 Product notation

This hasn't saved us any work yet, but we see, for example, that it would save us introducing a new letter to our alphabet in writing the 28th curve in a 28–3 configuration. Let's see how we can save ourselves some work. Consider, once again, the fifth curve A_5 in the 5–3 configuration. There are several things we may notice:

- The formula for A_5 involves several commutators on three letters, a typical one being $((A_1,A_2),A_3)$.

- The subscripts are in increasing order when we read from left to right.

- The several terms that appear consist of all commutators of the form $((A_i,A_j),A_k)$, with $i < j < k < 5$, where i, j and k are chosen from the integers 1, 2, 3 and 4.

If we look at other configurations, such as the 6–5 configuration, we would see a similar situation (when we use our subscript notation).

How can we incorporate these three observations in a single simple compact form? The notation below would follow standard mathematical notational procedures for complicated products. Write

$$A_5 = \prod_{i<j<k<5} ((A_i,A_j),A_k).$$

Let's dissect the notation:

- $((A_i,A_j),A_k)$ represents the typical term. For example, when $i = 1$, $j = 2$ and $k = 4$, we get $((A_1,A_2),A_4)$.

- The Greek letter Π (upper case π) indicates "product." We are not actually "multiplying" here, of course, but the notation we have been using all along suggests "multiplicative notation."

- Under the symbol Π we see $i < j < k < 5$. This indicates first that the subscripts that appear are in increasing order, and that all such subscripts with i, j, k integers greater or equal to 1 and less than 5 are included. Notice the "5" tells us that k is no larger than 4.

Example 4.8.1 How would this notation work for the construction of the 8–4 configuration? Let's do it in detail. We use our customary format with our new notation.

Begin with four separated curves A_1, A_2, A_3 and A_4. Add to this A_5 to get a 5–4 configuration, then A_6 to get a 6–4 configuration, then A_7 to get a 7–4 configuration and, finally, A_8 to get a 8–4 configuration. The formulas are evidently:

$$A_5 = (((A_1,A_2),A_3),A_4)$$

$$A_6 = \prod_{i<j<k<\ell<6} (((A_i,A_j),A_k),A_\ell).$$

$$A_7 = \prod_{i<j<k<\ell<7} (((A_i,A_j),A_k),A_\ell).$$

$$A_8 = \prod_{i<j<k<\ell<8} (((A_i,A_j),A_k),A_\ell).$$

◀

Simple, isn't it? Note we needed a fourth subscript, ℓ, here, because each commutator involved four *letters*. Note also that the *stopping point* (8 in our last formula) agrees with the subscript of the curve we are representing. What would be the formula for the eleventh curve of an 11–6 configuration built by our methods? The answer is simply

$$A_{11} = \prod_{i<j<k<\ell<m<n<11} (((((A_i,A_j),A_k),A_\ell),A_m),A_n).$$

Note that cutting away a single curve, say A_4, causes all commutators involving A_4 to reduce to 1, so what remains is equivalent to the tenth curve of a 10–6 configuration. When we have cut away all but 7 curves, we arrive at the 7–6 configuration, so cutting away one more curve, causes the remaining six to fall apart.

4.8.3   Subscripts on subscripts

Are we now, finally, finished with these linking problems? Almost, but not quite.

Once again, we are soon going to run out of letters—on the subscripts! For example, the 11–6 configuration involved the indices (i.e., letters or subscripts) i, j, k, ℓ, m, and n. If we wanted the 100th curve in the 100–70 configuration, we would need 70 letters. How can we modify our notation one more time to accommodate to such a configuration. As before, we introduce numerical subscripts on the *subscripts* themselves! Thus, in place of i, j, k, etc., we use i_1, i_2, i_3, etc. We can then write A_{11} in the 11–6 configuration using as indices i_1, i_2, i_3, ...i_6 in place of the more clumsy letters i, j, k, ℓ, m, n that we used previously:

$$A_{11} = \underset{i_1 < i_2 < i_3 < i_4 < i_5 < i_6 < 11}{(((((A_{i_1}, A_{i_2}), A_{i_3}), A_{i_4}), A_{i_5}), A_{i_6}).}$$

In practice we would prefer to omit some of the expressions by merely indicating with ellipses (i.e., three dots) that all these parentheses and subscripts are needed:

$$A_{11} = \underset{i_1 < i_2 < \cdots < i_6 < 11}{((\ldots((A_{i_1}, A_{i_2}), A_{i_3}), \ldots), A_{i_6}).}$$

Why all those dots? The first set of dots indicates that we haven't written in all of the parentheses: there should be five of them. We can reduce the mess by eliminating some and use the dots to indicate that more are really intended. The second set of dots, those under   , indicates that we have omitted the part of the expression

$$i_3 < i_4 < i_5$$

that should be included. Finally, the third set of dots, those inside the parentheses, indicates that we have not written in the elements A_{i_4} and A_{i_5}. The reader of such a formula is expected to understand what is missing and fill it in if necessary.

This convention saves us some writing once the pattern is clearly understood. For example, the 100–70 configuration would be almost impossibly complicated, but the dots help considerably. The 100th curve in that configuration is simply written as,

$$A_{100} : \underset{i_1 < i_2 < \cdots < i_{70} < 100}{((\ldots((A_{i_1}, A_{i_2}), A_{i_3}), \ldots), A_{i_{70}}).}$$

Problem 199 *Write the 50th curve in a 87–33 configuration in our new notation. (Assume the first 49 curves form a 49–33 configuration.)*

                                                                    Answer □

Problem 200 *How many winds are there in the eleventh curve in an 11–5 configuration?*

                                                                    Answer □

4.9   Groups

One of the many aspects of modern mathematics that distinguishes it from say, nineteenth-century mathematics, is that there is nowadays a good deal of emphasis on *abstract structures*. What this amounts to is that mathematicians will often study some abstract system defined axiomatically, which has, on the surface, no connection with the real world.

Why should one study such abstract systems? One reason is this. Many real life situations to which mathematics has been successfully applied appear to be quite different in nature, but actually involve the same mathematical analysis and the same structure. We saw this many times in the chapter on Nim. By studying the abstract structure, one reduces the analysis to its essentials. Anything one proves within the abstract setting then applies to each realization of the abstraction.

One such abstract structure is that of a *group*. We shall define the notion of a group and give a few examples of groups. We won't develop any of the theory of groups (there are many excellent books on groups) but the several examples will suggest how a general abstract theory could be useful in studying individual instances of the theory.

Definition of a group   A *group* is a set G together with an operation $\cdot$ that satisfies the following four[3] conditions:

1) If a and b are elements of G then $a \cdot b$ also belongs to G. (We might write ab for $a \cdot b$. Sometimes other notation such as $\times$ or $+$ is used for the operation to suggest multiplication or addition.)

2) There is an element 1 belonging to G such that $a \cdot 1 = 1 \cdot a = a$ for all a in G. The element 1 is called the identity. (Some times, when "+" is the notation for the operation, the identity is denoted by 0.)

3) If a is an element of G then there is an element a^{-1} called the inverse of a such that $a \cdot a^{-1} = 1$ and $a^{-1} \cdot a = 1$. (When $+$ is the symbol for the operation, one writes $-a$ in place of a^{-1}.)

4) If a, b, and c are elements of G then $(a \cdot b) \cdot c = a \cdot (b \cdot c)$. This is called the *associative law*.

At this point, the definition of a group is, of course, abstract. Let's look at some examples.

Example 4.9.1 Let G consist of the positive real numbers, and let "$\cdot$" denote usual multiplication (i.e., the operation that we would have written as "$\times$" in elementary school). Then

1. If a and b are positive real numbers, so is $a \cdot b$.

---

[3]Note that there is no fifth condition requiring that $a \cdot b = b \cdot a$. While many groups do have this property (the *commutative property*) we have seen that our group does not.

2. The usual number 1 satisfies $a \cdot 1 = a$ and $1 \cdot a = a$.

3. If a is a positive real number, then a^{-1} is denoted commonly as $1/a$ and $a^{-1} \cdot a = 1$ and $a \cdot a^{-1} = 1$. (a^{-1} is, of course, a real positive number if a is.)

4. This is just the usual associative law for the multiplication of real numbers.

How we group the numbers does not affect the outcome. For example,

$$4 \cdot (5 \cdot 6) = 4 \cdot 30 = 120$$

and

$$(4 \cdot 5) \cdot 6 = 20 \cdot 6 = 120.$$

◄

Example 4.9.2 Let G denote the integers (including the negative integers) with + for the operation. It is easy to see that this gives us a group. ◄

4.9.1 Rigid Motions

For those with a bit more background in mathematics, we mention that the examples that appeared in Section 4.2.1 (polynomials, functions, matrices, vectors) can all be endowed with a group structure by choosing some appropriate group operation. Other important groups involve symmetries, permutations, rotations, or rigid motions in a plane.

To illustrate, we can describe the appropriate group operation used in studying the group of rigid motions. It is similar to our operation in linking.

If A and B are rigid motions in a plane, then AB is just the the motion achieved by applying B then A. Suppose A consists of translating every point 2 units to the right, and B rotates a point 90 degrees about the origin. Then AB consists of first rotating, then translating, while BA consists of first translating, then rotating. In Figure 4.7 we show AB applied to a triangle T with one point at the origin.

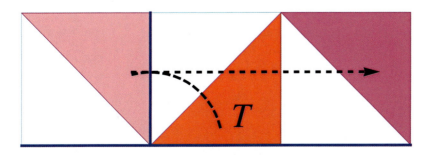

Figure 4.7: AB: First rotate the triangle T, then translate.

Figure 4.8 shows *BA* applied to the triangle *T*. (We see that the group of rigid motions is not commutative.)

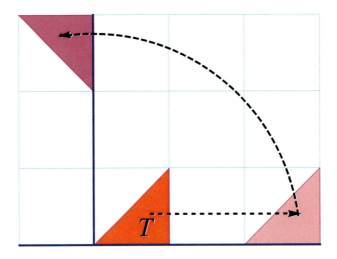

Figure 4.8: BA: First translate the triangle *T*, then rotate.

Note It is customary to read from right to left in describing rigid motions. Thus *AB* indicates doing *B* first.

4.9.2   The group of linking operations

Our algebraic work in this chapter can be viewed as part of group theory. We give an intuitive and mathematically incomplete description of this view now.

Suppose we begin, as we have many times, with three fixed separated circles *A*, *B*, and *C* in space. We view this as a starting set-up. Each starting set-up leads to its own linking group. (See Section 4.2.2.)

We consider all possible ways of weaving a fourth curve through *A*, *B*, and *C* and write down what we called our *formula for the fourth curve*. For example, ABC^{-1} would mean this: our curve goes through *A*, then *B* and then backwards through *C* and finally returns to its starting point. This *action* described by ABC^{-1} is an element of our group. So is any similar action.

What is our group operation that corresponds to the "·" we used in our definition of a group? It is simply *doing* one action then the other. For example, if we first *do* ABC^{-1} and then *do* $BAC^{-1}A$ we get

$$ABC^{-1}BAC^{-1}A$$

as

$$(ABC^{-1}) \cdot (BAC^{-1}A).$$

In this case, there are no simplifications possible. Sometimes, as we saw, there are simplifications possible. For example

$$(ABA^{-1}B^{-1}) \cdot (BAB^{-1}A^{-1})$$

reduces to 1. This follows from the associative law in the definition of a group. We leave it to the reader to verify that what we have been doing with links will satisfy the four stated conditions defining a group.

If we look at our first two examples of groups, involving ordinary numbers, we see that those groups are *commutative groups*, (e.g., $a \cdot b = b \cdot a$ and $a + b = b + a$ in the two examples). We have already observed that the groups involved with our linking problems are not always commutative. For example, if A and B are separated circles, then $AB \neq BA$. We can check this by using our model and verifying that $ABA^{-1}B^{-1} \neq 1$, i.e., it does not reduce to a curve that links neither A or B.

Problem 201 *Begin with two* linked *circles A and B. Wind a third curve through A and B according to the formula $ABA^{-1}B^{-1}$. Show that the third curve can be pulled free from A and B, thereby showing that $AB = BA$. (See Problem 188 in Section 4.5.)* □

Problem 201 illustrates one simple situation of a linking problem that gives rise to a commutative group. More generally, our groups are noncommutative, although a group might have special relations in it that do commutate. For example, if A and B are linked, but C is separated from A and B, then the relation $AB = BA$ is valid on the resulting group, but we can't write, for example, $AC = CA$. (Make a model.) Thus, for example, we can write $ABC = BAC$, but we can't write $ABC = ACB$ if we want to make a correct assertion in this group.

Problem 202 *Suppose that the circles A and B are linked, but C is separated from A and B. Simplify each of the following:*

$$ABA^{-1}B^{-1}A \quad and \quad ABCA^{-1}B^{-1}A.$$

Where feasible, check with a model. □

4.10 Summary and perspectives

There are a number of things to be learned from this chapter. Let's review what we did.

1. We started with two simple configurations (the 3–2 and 3–1 configurations) and then asked for the construction of a 4–3 configuration.

2. It became clear that pictures were inadequate for experimentation. So we made models which helped experimentation.

3. But the models soon proved inadequate. We came naturally to a method of keeping track of our actions: a bookkeeping system. This system soon began to look like ordinary algebra, although symbols such as *AB* did not mean multiplication of numbers.

4. We soon saw that this simple bookkeeping system was actually an algebraic system with very simple properties. And, more importantly, *it related directly to our linking problems.*

   Cutting away a curve corresponded to removing the corresponding letter. That little bit of algebraic structure helped enormously in solving the simpler linking problems. They were simpler because of our algebraic methods. The 5–3 configuration, for example, may have been impossibly difficult for us to construct without such a system.

5. The algebraic formulation actually helped pinpoint the intuitive idea of undoing everything we do, but to defer such undoings an appropriate length of time.

6. Even our algebra became prohibitive once we got to more complicated figures. The expressions just got too long. So we incorporated our intuition into the algebra and introduced commutators and compound commutators. The two useful notions involved "compounding" of commutators and "multiplying" commutators.

7. Even here, our notation was inadequate when we discussed configurations such as the 8–4 configurations. The notion of commutator simplified things, but still there were just too many commutators to combine.

8. So we finally obtained adequate notation in Section 4.8.1.

9. Note that our final notation does more than just shorten the writing of a formula. It contains the entire structure of any of our configurations. When we began the linking problems, we may have had no idea where we were heading. As we progressed we learned more and more about the structure of n–k configurations, and we incorporated what we learned in our notation. Section 4.8.1 developed the entire structure of such configurations, at least those constructed by our methods.

10. We also saw that some methods had the advantage of simplicity, but also had the disadvantage of not pointing us in the best direction for proceeding. For example, the fourth curve in the 4–2 configuration could be taken as

$$D = ABCA^{-1}B^{-1}C^{-1}.$$

This helped us obtain the nth curve in the n–2 configuration, but gave no hint for constructing the n–3 configurations. Our other method was less efficient, but did give us such a hint. This phenomenon often occurs in mathematics. One solves a problem, but the particular solution does not help us resolve new related problems[4]

11. Finally, we discussed groups *very* briefly. Groups provide a general framework for studying a number of mathematical systems such as those associated with linking problems. A good understanding of groups can help one understand other linking problems that we have not considered. And other problems not at all associated with links. Are groups the final answer? Of course not. Some algebraists study systems, very abstract systems, of which the entire theory of groups is just a simple example! And so it goes.

4.11 A Final Word

In a certain sense we have solved the linking problems we set out to solve. In another sense, we haven't.

Let's discuss this point, first in a broader context involving an aspect of evolution in mathematics and then in the specific context of what we have done with links. When a mathematical subject is in its infancy, it is not always entirely clear what, exactly, the subject under study is.

4.11.1 As mathematics develops

When earlier generations of mathematicians studied regions in space bounded by a surface, they might not have known exactly how *regions* or *surfaces* or *bounded* were to be interpreted. They had easy to visualize models in mind. For example *the region bounded by a sphere* makes sense (i.e., the inside of the sphere). So too does the inside of a bagel or pretzel. They also had no difficulty in counting the number of holes in a pretzel, even though the concept of *hole* may not have been well defined.

After all, if we look at a bagel we would agree that it has one hole, without our needing to know, in a strict mathematical sense, what a hole is. But, when one proceeds to more complicated surfaces, one needs to have a mathematically precise way of dealing with the concepts.

Earlier generations of mathematicians often made significant contributions to a subject even though some of their work was mathematically imprecise. We would say their work was not rigorous. One could say that their work contained errors. But, in a sense, it would be more descriptive to say their work contained *gaps*: their results were correct under somewhat more restrictive conditions than

---

[4]Recall that our solution for two, three, and four-marker games did not lead to a general solution until we completely changed our perspective and looked only at the gaps.

they supposed. For example, if one defines a region in a certain way and then *proves* a theorem about regions which is valid only if the region meets some extra conditions (not mentioned in the definition), one has a gap in reasoning.

Imprecise work is by no means worthless—it just doesn't always apply without some sort of modification. For example, one constantly applies plane geometry ideas in real life even though the surface of our earth is more like a sphere than like a plane. Thus we think of a baseball infield as planar square without that creating any real problems. The error in doing this is small, but would be great if the sides of our "square" were thousands of miles rather than 90 feet. Similarly, Newtonian physics is fine when it applies, but became inadequate to deal with such things as objects moving close to the speed of light or tiny objects such as atoms.

4.11.2   A gap?

After creating our algebraic structure, we assumed it faithfully expressed the structure of our linkings. It seemed to do just that. Actually it didn't!

Consider, for example, the expression ABA^{-1}. As an element of our group, it is not equal to the element B, because $ABA^{-1}B^{-1}$ is not equal to 1. But if our shoelace follows the instructions given by $C : ABA^{-1}$ applied to two separated circles, we find we can "slip" A off C, leaving B, as shown in Figure 4.9. Thus the formula ABA^{-1} does have the same linking properties as the formula B does. The algebra and the linking were at odds.

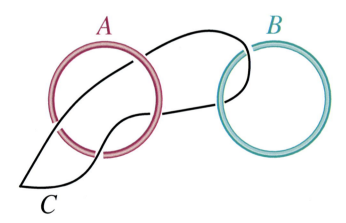

Figure 4.9: A "slips off" C.

The same thing would happen any time we had a formula of the form XYX^{-1}, where X and Y are any elements of our group. There would be a cancellation that takes place in the linking, but not in the group. In our group structure X and X^{-1} can't cancel each other (in XYX^{-1}) unless X and Y commute. But, when applied to links, such a cancellation can take place.

How can we take this into account in our work? Actually, we have taken it into account, although we have not stated this explicitly.

Our project was to construct $n - k$ configurations. We always had an initial set-up of k separated curves. All the formulas we have used for the shoelaces in our configurations are products of commutators or compound commutators. This problem of cancellation does not take place with commutators. For example, the shoelace for our 4-3 configuration had the formula

$$ABA^{-1}B^{-1}CBAB^{-1}A^{-1}C^{-1}.$$

No cancellation here!

We successfully avoided all cancellations in our development by restricting our group suitably to group elements that actually *do* describe the linking structure of n–k configurations. If we had tried to use the full group to represent all possible linking structures, we would have run into difficulties. As we saw with the curve ABA^{-1} applied to two separated curves A and B, the group would not describe some of the linkings accurately.

Example 4.11.1 Let's look at a rather artificial but clear example that illustrates this perspective. Suppose we wanted to prove that two squares in the plane are congruent if and only if their projections vertically onto the x-axis have the same length. We wouldn't be able to do that—the statement is false.

The group of rigid motions in the plane (see Section 4.9.1) includes rotations, and the length of a projection of a square can change under a rotation. But if we restrict our focus to only those squares that have horizontal and vertical sides, and we consider only the translations, the statement is true. It is clear that such a square can be translated onto another such square if and only if the two squares have the same side length, and that happens if and only if their projections onto the x-axis have the same length.

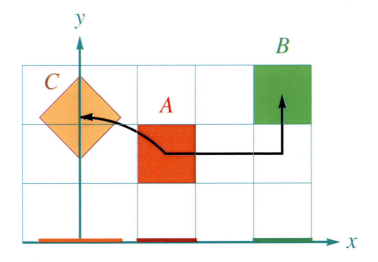

Figure 4.10: Projections of squares on the x-axis.

Note that in Figure 4.10, *A* can be translated onto *B*, but a rotation would be required to move *A* onto *C*. Here we have restricted our attention to only certain squares, and have used only those rigid motions that are translations. Rotations didn't enter the picture.   ◀

In our linking project, we didn't discuss all possible link structures; we restricted ourselves to separated curves and to *n–k* configurations. We used only the subgroup consisting of products of commutators and compound commutators. Other elements of the linking group played no role.

4.11.3   Is our linking language meaningful?

In our study of links, we used a number of terms that had intuitive content for all of us. Terms such as *curve, go through A backwards, winds, hangs together, falls apart,* etc. were certainly meaningful to us for purposes of communication, and even for purposes of making models of various of our configurations. We constantly used expressions such as *separated curves,* or *we can pull the curve out of A without cutting or tearing,* or *these two curves have the same linking properties.* We also sometimes gave warnings such as *be careful not to create any knots in your shoelace.* We did not, however, define these concepts in any precise way.

For our purposes it may not have been necessary to define all these terms. We can construct (at least theoretically) any configuration of the type we discussed. But we haven't solved any of these problems in a strict mathematical sense. The amount of mathematical machinery necessary even to discuss linking problems rigorously is enormous.

4.11.4   Avoid knots and twists

Let's illustrate the kinds of difficulties one encounters if one tries to mathematize our discussion.

1. We all know intuitively which direction is *backwards* when *we go through a curve C backwards.* Or do we? If *C* is a circle, we can all agree which direction is *forwards* and which direction is *backwards.* But can we agree which direction is which if *C* is the 50th curve in an 80–20 configuration?

2. One difficulty is that a curve *C* may have *twists* in it. Can we tolerate twists? Small twists, surely. But what about big twists? Or multiple twists? What exactly is a twist and what makes a twist *small, big,* or *multiple*? And are these distinctions really important? And are twists important? If so, can they be avoided?

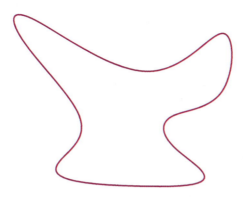

Figure 4.11: This curve can be transformed into a circle.

3. The curve in Figure 4.11 is not a circle, but can be transformed into a circle by *stretching*, *bending*, *pulling*, etc. No *cutting*, *tearing*, or *pasting* is necessary (whatever the precise meanings of these terms are).

4. The same is true of the curve C in Figure 4.12, although you may need to make a model to visualize it (simply untwist the ears to begin).

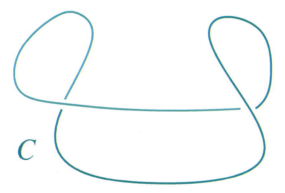

Figure 4.12: Curve with "ear-like" twists.

5. Now, add a circle A to the configuration passing through the ears to obtain the curve in Figure 4.13

Make a model and check that the resulting configuration will not come apart without cutting or tearing. But note that our curve with ears, C, *went through A and then through A backwards.*

In other words, our twisted curve has formula AA^{-1}. But it doesn't reduce to 1. What went wrong? In terms of our casual language, we allowed twists, not paying attention to our earlier warning to avoid knots and twists. But we still don't know what a twist is. If we remove the circle A, our *twisted curve* is not

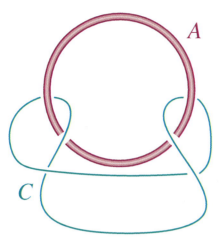

Figure 4.13: Is $C = AA^{-1}$?

distinguishable from any other curve according to our usage of the term curve. But, in the presence of A, its linking properties with A are quite different from the linking properties of an untwisted curve with formula AA^{-1}.

The point to these remarks is to make it clear that something we hadn't really come to grips with enters here. Our work isn't wasted. We can still construct our configurations if we can *avoid knots and twists* whatever that may mean. But, for a precise mathematical development, one would have to know what the terms mean and also how to deal with them mathematically.

4.11.5   Now what?

For our purposes it was easier simply to ignore the troublesome issues we have identified and possibly other issues we missed. If these can't be addressed, the status of our work is that we made a plausibility argument, but have not provided a rigorous proof. Maybe we (or someone else) can make our proof rigorous or find another proof. Or show that our result is false, perhaps by showing no configuration of some specific size is possible.

What is the current status of the problem?   The original work in this area dates back to paper by Hermann Brunn (1862–1939), in 1892. Brunn constructed n–$(n-1)$ configurations. He acknowledged that his work was not rigorous. The considerable technical machinery necessary to handle such problems rigorously had not yet been created. But, because of his original paper, today an n–k configuration is called an n–k *Brunnian link*.

In 1961, Hans Debrunner did provide a rigorous proof or Brunn's result. And, he rigorously proved the existence of all n–k configurations! Phew!

Then, in 1969, David Penney (see item [6] in our bibliography) provided a much simpler rigorous proof of the existence of all *n–k* configurations. Our chapter provides an intuitive, nonrigorous development leading to Penney's formulation. Penney's paper was only two pages long. It did not involve discovery of the solution. It used mathematical induction to verify that the solution via compound commutators works. He had to discover this somehow (perhaps along the lines of our development) but the actual paper was only a verification of a formula he had discovered.

This progression is common in mathematics. Someone discovers a result and proves it. Perhaps the proof is not rigorous. Someone else provides a rigorous proof. Then yet another mathematician finds a much simpler proof.

4.12 Answers to problems

Problem 165, page 184

You probably answered this without difficulty. You simply constructed three curves with each of the three pairs linked as shown in Figure 4.14 .

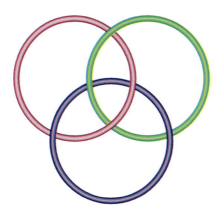

Figure 4.14: The three curves are linked in pairs.

Compare Figure 4.14 with the Borromean rings of Figure 4.1. Observe that, in Figure 4.1, the entire configuration is linked, but no pair of curves is linked. Here, each pair of curves is linked.

Figure 4.15 shows a different solution for this same problem made with a "shoelace" model. Comparing the solutions given in Figure 4.14 and Figure 4.15 we see that, had we chosen to make the shoelace go "backwards" through the second circle (instead of forwards as here) we would have constructed the same configuration in both. Should we have a language to describe backwards and forwards? (See Section 4.1.7 for the answer to this question.)

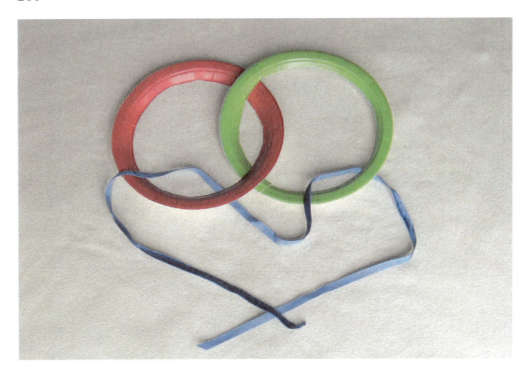

Figure 4.15: A shoelace model of a 3–1 configuration.

Problem 166, page 185

This is a bit harder than Problem 165 without cheating and looking at the Borromean rings for guidance, but you may have succeeded by reasoning more or less along the following lines. We may as well begin with two separated circles (as indicated below).

Figure 4.16: Start with two separated circles for Problem 166.

We now wish to "weave" a third curve through the two separated circles in such a way that the conditions of the problem are satisfied. That is our third curve (the "shoelace" if we made a model as suggested) must weave through the other two so that removal of one of the three curves causes the configuration to "fall apart." This must be true no matter which of the three curves is removed. It must also be true that the entire configuration of three curves "hangs together."

Now it is clear that, no matter how we weave in the "shoelace," removal of it will cause the other two curves to fall apart. (They are already separated.) Our task is to do the weaving in such a way that, if we removed either of the other two curves, the remaining one and the shoelace can be separated without cutting or tearing.

Once we understood this much, we could experiment with our shoelace and we might well arrive at a configuration such as the one in Figure 4.17.

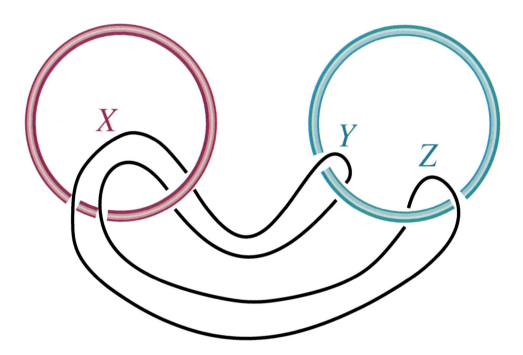

Figure 4.17: Weave the curve through the circles.

Let's see what happens with this configuration.

1. If we cut away the shoelace, the remaining two curves are already separated.

2. If, instead, we cut away the curve on the right as in Figure 4.18, the shoelace is draped over the remaining curve near the point X. If we hold the shoelace at X and pull, voilà, we have effected the separation.

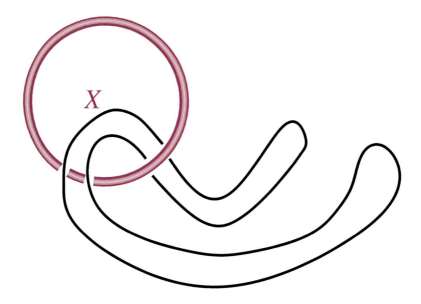

Figure 4.18: Cut away the circle on the right.

3. Similarly, if we cut away the curve on the left as shown in Figure 4.19, the shoelace is draped over the remaining curve in such a way that if we hold the shoelace with one hand near the point Y and with the other hand near the point Z and pull, once again, voilà!

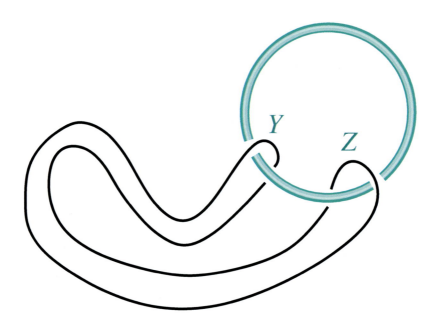

Figure 4.19: Cut away the circle on the left.

Can you visualize all that? Perhaps you need the model.

Problem 167, page 185

If you think it possible, look again at the reasoning that led to the construction of the 3–2 configuration to see if there is any discernible pattern that could be of some help. And use your model.

It is too hard to rely on a picture. It is much easier to experiment with three rings and a shoelace.

If, on the other hand, you think it is not possible, try to discover some basic irreconcilable difficulty (as you did with the tiling problems in Chapter 1).

Problem 168, page 187

Figure 4.20 illustrates[5] a curve (using the model) that is described by the expression

$$A^b BAB^b.$$

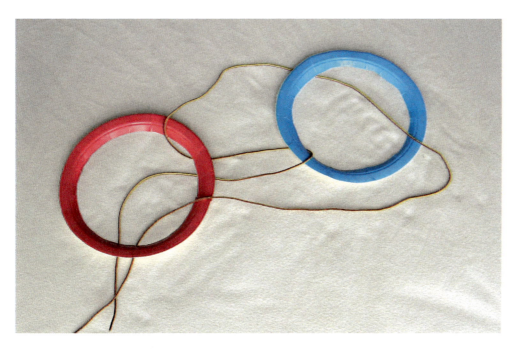

Figure 4.20: $A^b BAB^b$.

Figure 4.21 illustrates a curve described by the expression ABB^bA^b.

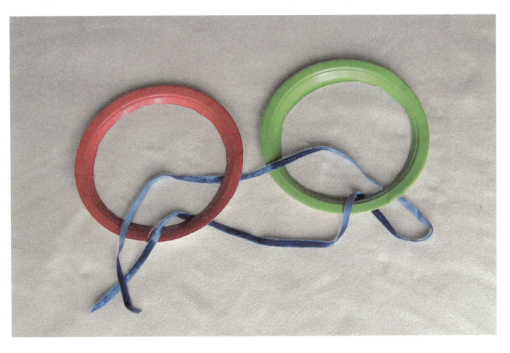

Figure 4.21: ABB^bA^b

Problem 169, page 187

The first two expressions BAB^bA^b and A^bBAB^b from Problem 168 give rise to 3–2 configurations. These are really essentially the same as the description we gave before with the roles of A and B reversed or the directions we chose as positive changed.

Problem 170, page 187

The last three expressions AA^bBB^b, AAA^bA^b, and ABB^bA^b result in configurations that can be separated with no more than a "pull." (No cutting or tearing necessary.)

Problem 171, page 187

When an action is undone immediately (as in AA^bBB^b, AAA^bA^b, and ABB^bA^b) it is as if the action had not been accomplished in the first place. Note how this can happen in stages. For example, in ABB^bA^b the element A is not undone until later, but B is undone at the first opportunity and that allows A to be undone too. This is necessarily vague (at this stage). Part of what we shall be doing is to make it more precise.

Problem 172, page 187

Yes. Taking the left ring as A and the right ring as B, the shoelace has the formula ABA^bB^b. So it is a 3–2 configuration.

Problem 173, page 188

If you have not constructed the 4–3 configuration yet, try again. Try to use what you have learned from Problem 171)

Problem 174, page 190

Check, using the ordinary rules of arithmetic, that each of the expressions
$$AA^b, \ BB^b, \ A^bA, \ B^bB, \ AB^bBA^b, \ \text{and} \ AAAA^bA^bA^b$$
is equal to 1.

Problem 175, page 190

Check, using the ordinary rules of arithmetic, that each of the expressions
$$A, \ ABB^b, \ AA^bA, \ \text{and} \ BAA^bB^bA$$
is the same as A.

Problem 177, page 192

Check that
$$ABC \Longrightarrow AC$$
$$ABCA^{-1}B^{-1}C^{-1} \Longrightarrow ACA^{-1}C^{-1}$$
$$ABA^{-1}B^{-1} \Longrightarrow 1$$
both algebraically and by thinking about what is really going on.

Problem 179, page 193

If, in the 4–3 configuration, we let X represent the curve
$$X = ABA^{-1}B^{-1}$$
then the 4–3 configuration takes the form
$$XCX^{-1}C^{-1}.$$

It's just like the 3–2 configuration except that X represents more than just a single link.

Problem 180, page 193

A construction is given in Section 4.4, but you should try before reading on. To do this construction you must first indicate the starting set-up, and then give an expression for the action of the fifth curve (the shoelace).

You will have to verify that no matter which of the five curves is cut away, the result is 1. Also, you should check that if no curve is cut away, the entire configuration hangs together. Unless you have a very good model, you will find it difficult actually to construct it physically.

Problem 182, page 196

The 6–5 configuration is constructed by beginning with 5 separated curves A, B, C, D, and E and winding the sixth curve through the given five according to the expression

$$((((A,B),C),D),E).$$

Note, as before, that replacing any of the letters A, B, C, D or E by 1 causes the entire expression to reduce to 1. Note also that the shoelace must go through 46 winds to complete its task.

Problem 183, page 196

That $((1,B),C) = 1$ follows from the reduction

$$((1,B),C) = (1,B)C(1,B)^{-1}C^{-1}$$

$$= 1B1^{-1}B^{-1}CB1B^{-1}1^{-1}C^{-1} = BB^{-1}CBB^{-1}C^{-1} = 1.$$

Here we have used the fact that $1^{-1} = 1$. Make sure you follow the above computations and fill in whatever steps are missing.

Problem 186, page 197

Problem 186 and Problem 187 are straightforward computations, but you may need to work out the details for Problem 188.

Problem 188, page 197

Suppose $XY = YX$. By definition

$$(X,Y) = XYX^{-1}Y^{-1}$$

and this is the same as $YXX^{-1}Y^{-1}$ because we are assuming for the problem that $XY = YX$. Thus, putting this all together, we have

$$(X,Y) = XYX^{-1}Y^{-1} = YXX^{-1}Y^{-1} = 1.$$

On the other hand, if $(X,Y) = 1$, then $XYX^{-1}Y^{-1} = 1$. Thus

$$\left(XYX^{-1}Y^{-1}\right)(YX) = 1(YX) = YX$$

while, it is also true that

$$\left(XYX^{-1}Y^{-1}\right)(YX) = XYX^{-1}Y^{-1}YX = XY.$$

Hence $XY = YX$. Make sure you understand each step of the argument above.

Problem 189, page 198

To answer this problem, we need to know what we are looking for. We will eventually have four curves A, B, C, and D. The breaking point for the 4–2 configuration is 2. Thus each pair of curves must be separated, but each group of three curves must hang together. Another way to say this is that each group of three curves must form a 3–2 configuration. For in that case, the three curves will hang together, but removal of one more curve will cause the two remaining curves to fall apart. Thus, the third curve C should form a 3–2 configuration with A and B. This gives rise to the formula (A,B) for C.

Problem 190, page 198

We first observe that the formula must have the quality that removal of a single letter leaves a 3–2 configuration. For in that case, removal of a second letter will cause the expression to collapse. Thus, you may expect there to be some symmetry in the roles of A, B, and C in the expression. Some students in the class suggested $ABCA^{-1}B^{-1}C^{-1}$ for the fourth curve D. Let us verify that this gives the desired result.

If we remove	We arrive at
A	$BCB^{-1}C^{-1}$
B	$ACA^{-1}C^{-1}$
C	$ABA^{-1}B^{-1}$

Each of these three resulting configurations is a 3–2 configuration. If we remove D, the curve we have just added, we arrive at the 3–2 configuration formed by A, B, and C. And that's just what we wanted.

Here is another expression you may have tried for D:

$$(A,B)(A,C)(B,C).$$

Why is that a natural expression to try? Well, we want removal of a single curve to result in a 3–2 configuration for the remaining curves. Algebraically, this is tantamount to the condition that removal of a single letter (more precisely, replacing the letter by 1), results in an expression for a 3–2 configuration. Let's check.

- If we replace A by 1:
$$(A,B)(A,C)(B,C) \Longrightarrow (1,B)(1,C)(B,C) = 11(B,C) = (B,C).$$

- If we replace B by 1: $(A,B)(A,C)(B,C) \Longrightarrow (A,C)$.

- If we replace C by 1: $(A,B)(A,C)(B,C) \Longrightarrow (A,B)$.

Make sure you understand these computations.

Problem 191, page 199

The first solution,
$$ABCA^{-1}B^{-1}C^{-1}$$
has six winds. The second $(A,B)(A,C)(B,C)$ has twelve winds. So the first is simpler and more efficient. And that's certainly a desirable quality.

Another desirable quality is that the construction give insights that are useful to further construction. For example, the 3–2 configuration gave insights to the 4–3 configuration which in turn helped us see how to construct the 5–4 configuration. Does either of the constructions of the 4–2 configuration help us see how to construct, for example, a 5–2 configuration? To find out the answer to this question, look at Problem 192–Problem 194 in Section 4.6.3.

Problem 192, page 199

Since the break-point of a 5–2 configuration is at "2" we began with two separated curves A and B. Following the reasoning of Section 4.6.2 we see that C should be added so that A, B, and C together form a 3–2 configuration. Then D should be added so that A, B, C, and D form a 4–2 configuration. Finally we add E. Let us see what our two solutions to the 4–2 configuration have to offer.

It is convenient to use a chart that gives complete directions for the construction. These directions should be such that a skilled worker who understands our notation would be able (at least in principle) to make a model. Begin with two separated curves, A and B. First attempt:

Add	Formula	Resulting Configuration
C	(A,B)	3–2
D	$ABCA^{-1}B^{-1}C^{-1}$	4–2
E	$ABCDA^{-1}B^{-1}C^{-1}D^{-1}$	?

Does the addition of E give rise to a 5–2 configuration? We need only check that removal of a single letter gives rise to a 4–2 configuration. For example, removal of A gives rise to the expression
$$BCDB^{-1}C^{-1}D^{-1}$$

so that the curves B, C, D, and E form a 4–2 configuration. Similar computations show that removal of any other single letter gives rise to a 4–2 configuration so the construction works. The curve E has only eight winds. Pretty efficient—we couldn't possibly get by with fewer. (Why?) What about our other construction of the 4–2 configuration? Does that give any insights? Let's try to see what is involved.

The expression for D in that construction was

$$(A,B)(A,C)(B,C).$$

This construction succeeded because removal of any of the three letters A, B, or C (more precisely, replacing the letter with 1) gave rise to a simple commutator. This represents a 3–2 configuration. We should observe that the expression was obtained by combining the three letters A, B, and C in pairs in all possible ways: A with B, A with C, and B with C. If we extended this idea, we would arrive at the following description of a configuration. Begin with two separated curves, A and B.

Add	Formula	Resulting Configuration
C	(A,B)	3–2
D	$(A,B)(A,C)(B,C)$	4–2
E	$(A,B)(A,C)(A,D)(B,C)(B,D)(C,D)$	?

Is this a 5–2 configuration? As before, we must show that replacing a single letter with 1 causes the expression to reduce to a 4–2 configuration. If, for example, we replace the letter A with 1, we arrive at

$$(1,B)(1,C)(1,D)(B,C)(B,D)(C,D),$$

which reduces to $(B,C)(B,D)(C,D)$, a 4–2 configuration formed by the remaining four curves B, C, D and E. A similar analysis shows that the same is true if we replace any of the other letters with 1: we always arrive at a 4–2 configuration for the remaining curves.

This solution is less efficient than the preceding one. In this case, the curve E has twenty-four winds. The last solution required only eight winds for E.

Problem 193, page 199

The 6–2 configuration involves no new ideas. We can extend either method that we have already used. For the first method, we would wind the sixth curve F through the 5-2 configuration already constructed according to the formula

$$ABCDEA^{-1}B^{-1}C^{-1}D^{-1}E^{-1}.$$

Ten winds in all.

Problem 194, page 199

The second method gives rise to this formula for F:

$$(A,B)(A,C)(A,D)(A,E)(B,C)(B,D)(B,E)(C,D)(C,E)(D,E).$$

Forty winds in all.

Problem 195, page 199

If you did the reasonable thing and tried to extend the efficient method described in Problem 189 and Problem 190 for the 4–2, 5–2, and 6–2 configurations, you probably ran into difficulty.

Basically, what allowed that method to be so efficient is that the formula for the 3–2 configuration involves only simple winds and inverses: we don't have to undo anything more complicated than a single wind. For example, in the expression

$$ABA^{-1}B^{-1},$$

A^{-1} undoes A, B^{-1} undoes B. Thus, the expression

$$ABCA^{-1}B^{-1}C^{-1}$$

allows removal of a single letter to result in a 3–2 configuration, as required in a is 4–2 configuration.

Now, with the 5–3 configuration, we are faced with something more complicated. Removal of a single curve must give rise to a 4–3 configuration. And in the 4–3 configuration some of the "undoings" undo commutators, not just simple winds. For example, in the expression

$$ABA^{-1}B^{-1}CBAB^{-1}A^{-1}C^{-1},$$

to undo C is simple, but to undo $ABA^{-1}B^{-1}$ requires the more complicated expression $BAB^{-1}A^{-1}$.

Perhaps you found a way of doing it. But does it offer any insights that will be useful in constructing more complicated configurations? Our second method was less efficient than our first in constructing configurations which had the breaking point at "2." But it did offer a clearer pattern for further construction.

For example, to construct the 5–2 configuration, the fifth curve, E followed a formula which played no favorites with respect to the letters A, B, C and D. It simply took all pairs of those four letters, formed the simple commutators on them, and followed one-after-another:

$$E: \quad (A,B)(A,C)(A,D)(B,C)(B,D)(C,D).$$

Removal of a single curve resulted in a 4–2 configuration as desired.

This suggests that the fifth curve of the 5–3 configuration could follow a formula which played no favorites with respect to the letters A, B, C and D, takes all triples of those letters, forms compound commutators on them, and follows one-after-another.

$$E: \quad ((A,B),C)((A,B),D)((A,C),D)(B,C),D).$$

Let's check. Removal of a single curve should result in a 4–3 configuration. If we remove A, for example, we arrive at $((B,C),D)$, which does represent a 4–3 configuration using the curves B, C, D and E.

The same result occurs, of course, if any other curve is removed. We shall not carry out a full computation here. We merely observe that, for example, $((1,B),C) = 1$ (see Problem 183).

Problem 198, page 200

The answer is given in Example 4.8.1.

Problem 199, page 203

$$A_{50}: \quad \underset{i_1 < i_2 < \cdots < i_{33} < 50}{} ((\ldots((A_{i_1},A_{i_2}),A_{i_3}),\ldots),A_{i_{33}}).$$

Problem 200, page 203

There are 11,592 winds in the eleventh curve in an 11–5 configuration. There are 252 commutators on 5 letters chosen from the 10 letters $A_1,\ldots,A_{10}$. Each such commutator has 46 winds, as we saw in Section 4.4.3.

.

Appendix A

Induction

The story is told[1] that when the great mathematician Karl Friedrich Gauss (1777–1855) was a child, his teacher asked the pupils to add up all the integers from 1 to 100, (perhaps as punishment for talking in class). Within a few seconds, Gauss came up with the answer, 5050.

Here is how Gauss achieved this so quickly, He reasoned as follows. Set up the sum

$$
\begin{array}{rcccccccc}
S & = & 1 & + & 2 & + & 3 & + & \ldots & + & 99 & + & 100 \\
S & = & 100 & + & 99 & + & 98 & + & \ldots & + & 2 & + & 1 \\
\hline
2S & = & 101 & + & 101 & + & 101 & + & \ldots & + & 101 & + & 101
\end{array}
$$

So twice the sum is 100×101 and the sum must be 5050.

The same technique could be used to show that for every positive integer

$$1+2+3+\ldots+n = \frac{n(n+1)}{2}. \tag{A.1}$$

Suppose, now, that we hadn't spotted this clever proof but nonetheless had begun to suspect some kind of formula would be true. We might experiment (with small values of n) as we did in all the problems we attacked, and guess the formula (A.1). We can then easily check the formula for $n = 1, 2, 3, \ldots$ up to quite large values. How far should we go in this process until we are convinced the formula is indeed true?

The answer is that no amount of checking constitutes to a proof for all values of n. A mathematical proof requires a verification for every value of n and checking a few million special cases does not prove the rest.

One way to verify that the formula works for all values of n uses the notion of *mathematical induction* which we discuss in this Appendix. We shall see that this technique is useful in many parts of mathematics, In fact, mathematical induction figures frequently in our problems dealing with Pick, Nim, and Links.

---

[1]The same story has been told about many different mathematicians. But it may be true in the case of Gauss.

A.1   Quitting smoking by the inductive method

Before applying induction to proving some mathematical statements let us try to get a sense of the method in an every-day setting. Suppose a person who wished to stop smoking knew that if he (or she) could stop for just one whole day, he could be sure to avoid smoking for the very next day. If that were true then, in fact, he would certainly stop smoking forever if only he can to stop *for one day*. This would get him started: each day that he did not smoke would lead to the next smoke-free day.

In connection with our formula (A.1) we could argue similarly. Suppose one can verify that (A.1) is valid for $n = 1$. (That's like being able to stop smoking for that one day). And suppose we could prove that *if the formula is true* for any particular positive integer n then it must be true for the next integer $n + 1$. (That's the analog of *knowing* that if he can go any full day without smoking, he can certainly go one more). If we can do that, then we will have proved the validity of (A.1) for all positive integers.

A.2   Proving a formula by induction

Let us return to the task of proving the formula that Euler discovered on his own.

$$1 + 2 + 3 + \cdots + (n - 1) + n = \frac{n(n+1)}{2}.$$

An easy *direct proof* of this would follow Euler's idea. Let S be the sum so that

$$S = 1 + 2 + 3 + \cdots + (n - 1) + n$$

or, expressed in the other order,

$$S = n + (n - 1) + (n - 2) + \cdots + 2 + 1.$$

Adding these two equations gives

$$2S = (n + 1) + (n + 1) + (n + 1) + \cdots + (n + 1) + (n + 1)$$

and hence

$$2S = n(n + 1)$$

or

$$S = \frac{n(n+1)}{2},$$

which is the formula we require.

Suppose instead that we had been unable to construct this proof. Lacking any better ideas we could just test it out for $n = 1, n = 2, n = 3, \ldots$ for as long as we had the patience. Eventually we might run into a counterexample (proving the theorem is false) or have an inspiration as to why it is true. Indeed we find

$$1 = \frac{1(1+1)}{2}$$

$$1 + 2 = \frac{2(2+1)}{2}$$

$$1 + 2 + 3 = \frac{3(3+1)}{2}$$

and we could go on for some time. On a computer we could rapidly check for several million values, each time finding that the formula is valid.

If the computer ever finds a counterexample (just one instance where the formula fails) then that would be a proof that it is a false formula.

But if the computer never finds a counterexample, if the formula proves to be correct after hours of checking? Is this a proof? If a formula works this well for untold millions of values of n, how can we conceive that it is false? We would certainly have strong emotional reasons for believing the formula if we have checked it for this many different values, but this would not be a mathematical proof.

Instead, here is a proof that uses the same method of induction that we had the smoker use to quit his habit.

Suppose that the formula does fail for some value of n. Then there must be a first occurrence of the failure, say for some integer N. We know $N \neq 1$ (since we already checked that) and so the previous integer $N - 1$ does allow a valid formula. It is the next one N that fails. But if we can show that this never happens (i.e., there is never a situation with $N - 1$ valid and N invalid), then we will have proved our formula.

For example, if the formula

$$1 + 2 + 3 + \cdots + M = \frac{M(M+1)}{2}$$

is valid, then

$$1 + 2 + 3 + \cdots + M + (M+1) = \frac{M(M+1)}{2} + (M+1)$$
$$= \frac{M(M+1) + 2(M+1)}{2} = \frac{(M+1)(M+2)}{2},$$

which is indeed the correct formula for $n = M + 1$. Thus there never can be a situation in which the formula is correct at some stage and fails at the next stage. It follows that the formula is always true. This is a proof by induction.

A.3   Setting up an induction proof

This may be used to try to prove any statement $P(n)$ about an integer n. We wish to prove that the statements

$$P(1), P(2), P(3), \ldots, P(n), \ldots$$

(all of them) are true.

Here are the steps:

Step 1 Verify the statement $P(n)$ for $n = 1$.

Step 2 (The induction step) Show that whenever the statement is true for any positive integer m it is necessarily also true for the next integer $m + 1$.

Step 3 Claim that the formula holds for all integers $n \geq 1$ by the principle of induction.

A.3.1   Starting the induction somewhere else

An inductive argument is, on occasion, somewhat more convenient if the statements are labeled differently. Thus instead of wanting to prove the statements

$$P(1), P(2), P(3), \ldots,$$

we might want to prove the statements

$$P(0), P(1), P(2), P(3), \ldots$$

or even,

$$P(3), P(4), P(5), \ldots.$$

There is nothing new here, just a different use of labels. Induction proceeds in the same way. For example here is the scheme that we would use to prove that each of the statements

$$P(0), P(1), P(2), P(3), \ldots$$

is true.

Step 1 Verify the statement $P(n)$ for $n = 0$.

Step 2 (The induction step) Show that whenever the statement is true for any integer $m \geq 0$ it is necessarily also true for the next integer $m + 1$.

Step 3 Claim that the formula holds for all $n \geq 0$ by the principle of induction.

A.3.2   Setting up an induction proof (alternative method)

This alternative format may also be used to try to prove any statement $P(n)$ made about an integer n. In this version we do not go from step n to step $n + 1$.

Instead we may rely upon *any or all of the steps* from 1, 2, ..., up to n itself to help verify step $n + 1$. As before, we wish to prove that the statements

$$P(1), P(2), P(3), \ldots, P(n), \ldots$$

(all of them) are true.

Here are the steps:

Step 1 Verify the statement $P(n)$ for $n = 1$.

Step 2 (The induction step) Show that whenever the statement is true for *all* positive integers 1, 2, ..., m it is necessarily also true for the next integer $m + 1$.

Step 3 Claim that the formula holds for all $n \geq 1$ by the principle of induction.

Note that the induction step is different in this method. Whereas before we assumed that $P(m)$ was true and fashioned a proof that $P(m+1)$ should then be true, here we assumed more. We assumed that all of the statements

$$P(1), P(2), P(3), \ldots, P(m)$$

are true, and then we found a proof that $P(m+1$ should be true.

In the exercises you are asked for induction proofs of various statements. You might try too to give direct (noninductive) proofs. Which method do you prefer?

Problem 203 *For every positive integer n, $2^n > n$.* □

Problem 204 *Formulate the example of the person who wished to give up smoking in the language of Mathematical Induction. That is, what are the statements $P(n)$ for $n = 1, 2, 3, \ldots$?* □

Problem 205 *Prove by induction that for every $n = 1, 2, 3, \ldots$,*

$$1^2 + 2^2 + 3^2 + \cdots + n^2 = \frac{n(n+1)(2n+1)}{6}.$$

Answer □

Problem 206 *Compute for $n = 1, 2, 3, 4$ and 5 the value of*

$$1 + 3 + 5 + \cdots + (2n - 1).$$

This should be enough values to suggest a correct formula. Verify it by induction. □

Problem 207 *Prove by induction for every $n = 1, 2, 3, \ldots$ that the number*

$$7^n - 4^n$$

is divisible by 3. □

Problem 208 *Prove by induction that for every* $n = 1, 2, 3, \ldots$

$$(1+x)^n \geq 1 + nx$$

for any $x > 0$.   □

Problem 209 *Prove by induction that for every* $n = 1, 2, 3, \ldots$

$$1 + r + r^2 + \cdots + r^n = \frac{1 - r^{n+1}}{1 - r}$$

for any real number $r \neq 1$.   □

Problem 210 *Prove by induction for every* $n = 1, 2, 3, \ldots$ *that*

$$1^3 + 2^3 + 3^3 + \cdots + n^3 = (1 + 2 + 3 + \cdots + n)^2.$$

□

Problem 211 *Show that the following two principles are equivalent (i.e., assuming the validity of either one of them, prove the other).*

> **(Principle of Induction)** *Let S be any set of positive integers such that:*
>
> *1.* 1 *belongs to S.*
>
> *2.* *For all integers n, if n is is in S, then so is* $n + 1$.
>
> *Then S contains every positive integer.*

and

> **(Well Ordering of** $\mathbb{N}$**)** *If S is a set of positive integers and contains at least one element, then S has a first element (i.e., a minimal element).*

□

Problem 212 (Birds of a feather flock together) *Any collection of n birds must be all of the same species.*
Proof. This is certainly true if $n = 1$. Suppose it is true for some value n. Take a collection of $n + 1$ birds. Remove one bird and keep him in your hand. The remaining birds are all of the same species. What about the one in your hand? Take a different one out and replace the one in your hand. Since he now is in a collection of n birds he must be the same species too. Thus all birds in the collection of $n + 1$ birds are of the same species. The statement is now proved by induction. *[Criticize this "proof."]*   Answer □

Problem 213 *There were many possible uses of induction in Nim. For example, in 2-pile Nim we asserted that positions of the form* (n, n) *were all balanced. Give an inductive proof of this fact.*   Answer □

Problem 214 *Use induction and the fact that the Nim games* (k,k) *are balanced to prove that the 3-pile Nim game of the form* $(1,b,c)$ *is balanced if b is even and* $c = b+1$. □

Problem 215 *The inequality*
$$2^{n+1} > 2^n + 2^{n-1} + \cdots + 2^1 + 2^0$$
can be used in our analysis of the game of Nim. Verify it by induction.

Answer □

Problem 216 *Prove using induction: For every positive integer n,*
$$2^{n+1} - 1 = 2^n + 2^{n-1} + \cdots + 2 + 1.$$
(Thus, for example $2^4 - 1 = 2^3 + 2^2 + 2 + 1$.)

The problem shows that the largest binary numeral with a fixed number of bits is one less than the smallest having one more bit, e.g.,
$$10000 - 1 = 1111 \ (base \ 2).$$

□

Problem 217 *What is the corresponding statement for base ten of the statement in Problem 216 ?* □

A.4 Answers to problems

Problem 205, page 233

Check for $n = 1$. Assume that
$$1^2 + 2^2 + 3^2 + \cdots + n^2 = \frac{n(n+1)(2n+1)}{6}$$
is true for some fixed value of n. Using this assumption (called the induction hypothesis in this kind of proof), try to find an expression for
$$1^2 + 2^2 + 3^2 + \cdots + n^2 + (n+1)^2.$$
It should turn out to be exactly the correct formula for the sum of the first $n+1$ squares. Then claim the formula is now proved for all n by induction.

Problem 212, page 234

The induction step requires us to show that if the statement for n is true, then so is the statement for $n+1$. This induction step must be true if $n = 1$ and if $n = 2$ and if $n = 3 \ldots$, in short, for all n. Check the induction step for $n = 3$ and you will find that it does work; there is no flaw. Check the induction step for $n = 4$ and again you will find that it does work.

But does it work for all $n \geq 1$? Well yes and no. Yes for $n = 3, n = 4, n = 5$, ..., but no for $n = 1$ and $n = 2$.

Problem 213, page 234

For each integer $n = 0, 1, 2, 3, \ldots$ we let $P(n)$ be the statement that the position (n, n) is balanced in a 2-pile Nim game. The induction starts at $n = 0$ and the needed steps are:

1. We prove $P(0)$ is true.

2. We prove that if $P(k)$ is true whenever $0 \leq k \leq n$, then $P(n+1)$ is true.

Then we know, by induction, that $P(n)$ is true for all integers $n \geq 0$.

Here it is convenient to begin our induction at $n = 0$. Now, $P(0)$ is the statement that the position $(0, 0)$ is balanced. This is true since the final position in a Nim game is always balanced.

To verify the induction step, suppose that $P(k)$ is true whenever $0 \leq k \leq n$. Consider the game $(n+1, n+1)$. Any move from this position results in a game $(m, n+1)$ or $(n+1, m)$ where $m < n+1$. A balancing response resulting in the position (m, m) is possible since $m < n+1$. This position is balanced (by the Induction Hypothesis). Thus $(n+1, n+1)$ is balanced; i.e. $P(n+1)$ is true, as was to be proved.

You may have noticed that we needed the full Induction Hypothesis that $P(k)$ is true *for all $k < n$* in order to verify that $P(n+1)$ is true. In many of the other applications of an inductive argument it was enough to assume only that $P(n)$ was true in order to prove that $P(n+1)$ is true.

Problem 215, page 235

To do this, let $P(k)$ be the statement:
$$P(k): \quad 2^{k+1} > 2^k + 2^{k-1} + \cdots + 2^1 + 2^0.$$
Our first step is to show that
$$P(1): \quad 2^2 > 2^1 + 2^0$$
is true. But this amounts only to checking that $4 > 3$.

Suppose now (the induction hypothesis) that $P(n)$ is true. Thus (with n some fixed positive integer) we are assuming that
$$P(n): \quad 2^{n+1} > 2^n + 2^{n-1} + \cdots + 2^1 + 2^0$$
is a true statement.

We wish to show that the statement $P(n+1)$ is true, i.e., our goal is to prove that
$$P(n+1): \quad 2^{[n+1]+1} > 2^{[n+1]} + 2^{[n+1]-1} + \cdots + 2^1 + 2^0$$

is a true statement. We can do this by using our induction hypothesis.

Check that

$$2^{n+2} = 2(2^{n+1}) > 2(2^n + \cdots + 2^1 + 2^0),$$

because of the induction hypothesis. Note that both sides of the inequality are even numbers. It follows that

$$2^{n+2} > 2^{n+1} + \cdots + 2^2 + 2^1 + 2^0.$$

This is exactly the statement $P(n+1)$ and so that statement is true. The inequality now follows by induction for all values of $n = 1, 2, 3, \ldots$.

Appendix B

Nim, A Game with a Complete Mathematical Theory

Charles L. Bouton, *Nim, A Game with a Complete Mathematical Theory,* Annals of Mathematics, Second Series, Vol. 3, No. 1/4 (1901–1902), pp. 35–39.

NIM, A GAME WITH A COMPLETE MATHEMATICAL THEORY.

BY CHARLES L. BOUTON.

THE game here discussed has interested the writer on account of its seeming complexity, and its extremely simple and complete mathematical theory.[*] The writer has not been able to discover much concerning its history, although certain forms of it seem to be played at a number of American colleges, and at some of the American fairs. It has been called Fan-Tan, but as it is not the Chinese game of that name, the name in the title is proposed for it.

1. Description of the Game. The game is played by two players, *A* and *B*. Upon a table are placed three piles of objects of any kind, let us say counters. The number in each pile is quite arbitrary, except that it is well to agree that no two piles shall be equal at the beginning. A play is made as follows:—The player selects one of the piles, and from it takes as many counters as he chooses; one, two, . . ., or the whole pile. The only essential things about a play are that the counters shall be taken from a single pile, and that at least one shall be taken. The players play alternately, and the player who takes up the last counter or counters from the table wins.

It is the writer's purpose to prove that if one of the players, say *A*, can leave one of a certain set of numbers upon the table, and after that plays without mistake, the other player, *B*, *cannot* win. Such a set of numbers will be called a *safe combination*. In outline the proof consists in showing that if *A* leaves a safe combination on the table, *B* at his next move cannot leave a safe combination, and whatever *B* may draw, *A* at his next move can again leave a safe combination. The piles are then reduced, *A* always leaving a safe combination, and *B* never doing so, and *A* must eventually take the last counter (or counters).

2. Its Theory. A *safe combination* is determined as follows: Write the number of the counters in each pile in the binary scale of notation,[†] and

---

[*] The modification of the game given in §6 was described to the writer by Mr. Paul E. More in October, 1899. Mr. More at the same time gave a method of play which, although expressed in a different form, is really the same as that used here, but he could give no proof of his rule.

[†] For example, the number 9, written in this notation, will appear as
$$1 \cdot 2^3 + 0 \cdot 2^2 + 0 \cdot 2^1 + 1 \cdot 2^0 = 1001.$$

36   BOUTON.

place these numbers in three horizontal lines so that the units are in the same vertical column.   If then the sum of *each* column is 2 or 0 (*i. e.* congruent to 0, mod. 2), the set of numbers forms a safe combination.   For example,

$$1\ 0\ 0\ 1,$$
$$1\ 0\ 1,$$
$$1\ 1\ 0\ 0,$$

or 9, 5, 12 is a safe combination.   It is seen at once that if any two numbers be given, a third is always uniquely determined which forms a safe combination with the two given numbers.   Moreover, it is obvious that if *a*, *b*, *c* form a safe combination any two of the numbers determine the remaining one, that is, the system is closed.   A particular safe combination which is used later is that in which two piles are equal and the third is zero.   In the proofs which follow, the binary scale of notation is used throughout.

THEOREM I.   *If A leaves a safe combination on the table, B cannot leave a safe combination on the table at his next move.*   B can change only *one* pile, and he must change one.   Since when the numbers in two of the piles are given the third is uniquely determined, and since *A* left the number so determined in the third pile (*i. e.*, the pile from which *B* draws) *B* cannot leave that number. Hence *B* cannot leave a safe combination.

THEOREM II.   *If A leaves a safe combination on the table, and B diminishes one of the piles, A can always diminish one of the two remaining piles, and leave a safe combination.*   Consider first an example.   Suppose *A* leaves the safe combination *nine, five, twelve,* and that *B* draws *two* from the first pile, leaving the numbers *seven, five, twelve,* or

$$1\ 1\ 1,$$
$$1\ 0\ 1,$$
$$1\ 1\ 0\ 0.$$

If *A* is to leave a safe combination by *diminishing* one of the piles, it is clear that he must select the third pile, that containing *twelve*.   The number which is safe with 111 and 101 is 10, or *two*.   Hence *A* must leave *two* in the pile which contains *twelve*, or draw *ten* from that pile, and by doing so he leaves a safe combination.

To prove the general theorem, let the numbers, expressed in the binary scale, be written with the units in a vertical column, and suppose that *A* left a safe combination.   *B* selects one of the piles and diminishes it.   When a number of the binary scale is diminished it is essential to notice that in going over the number from left to right the first change which occurs is that some 1 is changed to 0, for if a 0 were changed to 1 the number would be increased whatever changes were made in the subsequent digits.* Consider, then, this first column, counting from the left, in which a change occurs. One and only one of the other two numbers will contain 1 in this same column, for A left a safe combination. Let A select the pile which contains the 1 in this column, and change the number by writing 0 in this column, and filling the remaining columns to the right with 0 or 1 so as to make a safe combination. The columns to the left remain unchanged, since they already have the required form. The new number so formed will be *less* than that in the pile which A selected. Hence whatever B draws, A can always diminish one of the piles, and leave a safe combination. That is, if A at any play can leave a safe combination on the table, he can do so at every subsequent play, and B never can do so.

If the play continues in this way A must win. For one of the piles must be reduced to zero by either A or B. If B reduces it to zero, the two remaining piles will be unequal, since B can never leave a safe combination, and A at his next move will make them equal, and will thereafter always leave them equal. B must, therefore, reduce the second pile to zero, and A then takes all of the third pile, and wins. If, on the other hand, A is the first player to reduce one of the piles to zero, he leaves the other two piles equal and wins as before. Hence we see that the player who can first leave a safe combination on the table should win.

If it happens that in the beginning a safe combination is placed on the table, the second player should win. If in the beginning a safe combination is not placed on the table, it is easily seen that the first player can always leave a safe combination by diminishing some one of the piles, and he can often do this by drawing from either one of the three piles. Therefore in this case the first player should win. That is, the first player should win or lose according as a safe combination is not or is placed on the table at the beginning.

3. The Chance of a Safe Combination. Assuming that the number in each pile at the beginning was determined by chance, let us compute the chance of a safe combination's being placed upon the table. It is easily shown that if each pile contains less than 2^n counters and if no pile is zero (*i. e.* if there are three piles), the possible number of different piles is

---

* The proof of this statement depends on the fact that the number 100 . . . (n ciphers), or 2^n, is greater than the number 11 . . . (n ones), or $2^{n-1} + 2^{n-2} + . . + 2 + 1 = 2^n - 1$.

38   BOUTON.

$$\frac{2^{n-1}(2^{2n}-1)}{3}.$$

The number of safe combinations in the same case is

$$\frac{(2^{n-1}-1)(2^n-1)}{3}$$

Hence the chance of a safe combination's being placed upon the table at first is

$$\frac{2^{n-1}-1}{2^{n-1}(2^n+1)},$$

and this is the chance that the second player should win. The chances of the first player's winning are to those of the second as

$$2^n + 2 + \frac{3}{2^{n-1}-1} \quad \text{to} \quad 1,$$

on the assumption that both players know the theory, and that the numbers in the various piles were determined by chance.

   4. A List of Safe Combinations, $n = 4$. The following are the 35 safe combinations all of whose piles are less than 16:

1	2	3	2	4	6	3	4	7	4	8	12
1	4	5	2	5	7	3	5	6	4	9	13
1	6	7	2	8	10	3	8	11	4	10	14
1	8	9	2	9	11	3	9	10	4	11	15
1	10	11	2	12	14	3	12	15			
1	12	13	2	13	15	3	13	14			
1	14	15									

5	8	13	6	8	14	7	8	15
5	9	12	6	9	15	7	9	14
5	10	15	6	10	12	7	10	13
5	11	14	6	11	13	7	11	12

Of course, to give *all* safe combinations of numbers less than 16 we should have to add to the above table the 15 of the form 0, n, n.

   5. Generalization. The foregoing game can be at once generalized to the case of any number of piles, with the same rule for playing. In this case a safe combination is a set of numbers such that, when written in the binary scale and arranged with the units in the same vertical column, the sum of each column is even (*i. e.*, $\equiv 0$, mod. 2). Just as before, it is shown that the player who first leaves a safe combination can do so at every subsequent play, and will win. The induction proof is so direct that it seems unnecessary to give it.

6. Modification. The game may be modified by agreeing that the player who takes the last counter from the table *loses*. This modification of the three pile game seems to be more widely known than that first described, but its theory is not quite so simple.

A safe combination is defined just as in the first case, *except* that 1, 1, 0 is *not* a safe combination, and 1, 1, 1 and 1, 0, 0 *are* safe combinations. When the first theory indicates that A should play 1, 1, 0 he must play either 1, 1, 1 or 1, 0, 0. The earlier part of the proof proceeds as before. In order to complete it, we must show that B can never leave 1, 1, 1; that, when 1, 1, 0 is indicated for A, he can always play either 1, 0, 0 or 1, 1, 1; and finally that, if the play is carried out in this way, B must take the last counter. That B can never leave 1, 1, 1 is at once clear, for A never leaves 1, 1, n where $n > 1$, since this is not a safe combination. Secondly, let us consider what sets of numbers B can leave which would indicate 1, 1, 0 as A's next play in the first form of game. They are 1, 1, n where $n > 1$, and 1, n, 0 where $n > 1$. In the first case A leaves 1, 1, 1 and in the second 1, 0, 0. The proof is now easily completed. Either A or B reduces a pile to zero. If B does so, the other two piles are unequal and both greater than unity, or at least one of the two remaining piles is unity. In the latter case A obviously wins. In the former case A makes the two piles equal, and then keeps them equal until B reduces one of them to 1 or 0. If B makes it 1, A takes all the other pile; if B makes it 0, A takes all but 1 of the other pile. Hence if B first reduces a pile to zero A wins. If A first reduces a pile to zero he leaves the other two piles equal and each greater than unity, and wins as before. Hence if A plays on the safe combinations as here modified, B must take the last counter from the table, and *loses*. That is, in this modified game, also, the player who can first get a safe combination should win.

This modified game can also be generalized to any number of piles. The safe combinations are the same as before, *except* that an odd number of piles, each containing one, is now safe, while an even number of ones is not safe.

HARVARD UNIVERSITY,
    CAMBRIDGE, MASSACHUSETTS.

Bibliography

[1] Anatole Beck, Michael N. Bleicher, and Donald Warren Crowe, *Excursion into Mathematics: The Millennium Edition*. With a foreword by Martin Gardner. A. K. Peters, Ltd. (Natick, MA: 1969, 2000). ISBN 1-56881-115-2 136

[2] Charles L. Bouton, *Nim, A Game with a Complete Mathematical Theory*, Annals of Mathematics, Second Series, Vol. 3, No. 1/4 (1901–1902), pp. 35–39. [A reproduction of the original paper appears in our Appendix.]

[3] R. L. Brooks, C. A. B. Smith, A. H. Stone and W.T. Tutte, *The dissection of rectangles into squares,* Duke Math. J. (1940) 7 (1): 312–340.

[4] W. W. Funkenbusch, *From Euler's Formula to Pick's Formula using an Edge Theorem*, The American Mathematical Monthly, Volume 81 (1974) pp. 647–648. 51

[5] Branko Grünbaum and G. C. Shephard, *Pick's Theorem*, The American Mathematical Monthly, Volume 100 (1993) pages 150-161. 51

[6] David E. Penney, *Generalized Brunnian Links*, Duke Math. J. (1969) 31–32. 214

[7] Georg Pick, *Geometrisches zur Zahlenlehre* Sitzungber. Lotos, Naturwissen Zeitschrift Prague, Volume 19 (1899) pages 311-319. 51

[8] J. E. Reeve, *On the Volume of Lattice Polyhedra*, Proceedings of the London Mathematical Society, 1957, 378–395. 91

[9] Sherman K. Stein, *Mathematics, The Man Made Universe*, Dover Publications, 3rd Edition (November 18, 2010). 14, 15

[10] Hugo Steinhaus, *Mathematical Snapshots*, 3rd ed. New York: Dover, pp. 266–267, 1999.

[11] W. T. Tutte, *The dissection of equilateral triangles into equilateral triangles*, Proc. Cambridge Phil. Soc., 44 (1948) 463–482. 15

[12] W. T. Tutte, *Squaring the Square*, Chapter 17 in Martin Gardner's *The 2nd Scientific American book of mathematical puzzles and diversions*, Simon and Schuster, 1961. 11

[13] W. T. Tutte, *Graph Theory As I Have Known It*, Clarendon Press, Oxford Lecture Series in Mathematics and Its Applications, 1998. 16

[14] Dale E. Varberg, *Pick's Theorem Revisited*, The American Mathematical Monthly Volume 92 (1985) pages 584-587. 51

[15] Edwin Buchman, *The Impossibility of Tiling a Convex Region with Unequal Equilateral Triangles* The American Mathematical Monthly, Vol. 88, No. 10 (Dec., 1981), pp. 748–753 15

[16] *The mathematical Gardner*. Edited by David A. Klarner. Wadsworth International, Belmont, Calif.; PWS Publishers, Boston, Mass., 1981. viii+382 pp. ISBN: 0-534-98015-5 15

[17] M. Ram Murty and Nithum Thain, *Pick's theorem via Minkowski's theorem*. Amer. Math. Monthly 114 (2007), no. 8, 732–736. 50

Index

Made in the USA
Las Vegas, NV
28 May 2024

90468189R00162